Painting and Decorating

Painting and Decorating

Roy Hughes

ELSEVIER

AMSTERDAM • BOSTON • HEIDELBERG • LONDON
NEW YORK • OXFORD • PARIS • SAN DIEGO
SAN FRANCISCO • SINGAPORE • SYDNEY • TOKYO

Newnes is an imprint of Elsevier

Newnes

Newnes is an imprint of Elsevier
Linacre House, Jordan Hill, Oxford OX2 8DP, UK
30 Corporate Drive, Suite 400, Burlington, MA 01803, USA

First edition 2005
Reprinted 2008

British Library Cataloguing in Publication Data
A catalogue record for this book is available from the British Library

Library of Congress Cataloging-in-Publication Data
A catalog record for this book is available from the Library of Congress

ISBN: 978-0-7506-6737-1

For information on all Newnes publications
visit our website at www.newnespress.com

Printed and bound in *China*

08 09 10 10 9 8 7 6 5 4 3 2

Contents

Preface   vii

1 Tools and equipment   1
2 Preparation of surfaces   17
3 Preparation of materials   59
4 Preparation of the work area   71
5 Brushes and rollers   91
6 Basic coatings (paints, stains and varnishes)   101
7 Application of coatings   115
8 Application of wall and ceiling hangings   127
9 Colour in decoration   153
10 Decorative paint effects   163
11 Scaffold   183
12 Health and safety   193
13 Glossary   205
14 Self assessment model answers   209

Colour plate section   227
Index   231

Preface

The information contained in this publication is aimed at practising painters and decorators or those entering the profession as modern apprentices. It will also be a useful source of reference for any person who may be in the process of decorating their properties.

It contains not just trade information but includes technical information that considers the actions required before, during and after the decoration has taken place. A knowledge of surfaces is required to determine the preparation required. An awareness of tools and materials is required to enable selection for the implementation of the decorating tasks.

Without prejudice I believe the content to be correct and this content will help the reader to progress to a technical certificate and eventually a NVQ award once site experience has been gained. The level of the book is at NVQ2 but there are parts that include L3 information and that will be useful at a later time.

I have not included specialist information on optional units such as spraying or application of decorative plasterwork, nor have I linked the chapters to units and elements of NVQ frameworks.

This book mirrors to some extent a publication produced by the construction industry training board, which is a manual of teaching and learning material for prospective colleges and training agencies delivering painting and decorating programmes, to use and adapt as they wish.

However not all people can attend a training programme whilst working, so enclosed throughout the book are working sections where you can test your knowledge and understanding, and collect information to use as evidence in gaining a NVQ award or technical certificate.

My thanks go to all the people and organisations I have worked with over the years whose own contributions have added to my own in assembling such knowledge that I hope are in chunks of comprehensible learning material.

1 Tools and equipment

1.1 Lists and tasks: Introduction

In the painting and decorating profession many tasks are carried out that require the use of a large selection of tools and equipment.

The painting and decorating of a task is not the only thing to take into consideration; in many instances the area to be decorated has to be assessed to determine if the proposed decoration can be carried out. It may be necessary to carry out some remedial work such as the replacing of damaged cornice, the repair of mouldings or the replacement of sections of skirting. To this end the decorator is required to carry out this work as other tradespersons will not give up their time for such a small task.

To carry out such tasks requires the decorator to have available a comprehensive service and maintenance tool kit in addition to the painter and decorator tool kit.

1.1.1 Hand tools

Tools and Equipment are used for the preparation of:

- surfaces and materials
- the application of materials.

Preparation tasks include the:

- removal of paint coatings, varnishes and lacquers
- removal of surface contaminants
- removal of wall coverings
- making good of surfaces by using fillers and stoppers
- abrading of surfaces before the application of products
- preparation of materials for use in application to surface areas.

The main application tasks are painting and decorating. The following hand tools are used for painting and decorating tasks:

- broad knife or scraper
- chisel knife
- shave hook
- paint stirrer
- filling knife
- putty or stopping knife
- palette knife.

The following are tools and items used for general tasks:

- hammers/ball/claw/pin
- pincers/pliers
- screwdrivers, various types
- caulking tool
- wire brush
- nail block
- sweeping brushes/pans/shovel
- nail punch
- wood chisels
- pointing trowel
- filling board or hawk
- rubbing block
- buckets
- dust brush.

All of the above-listed tools and items make up a good tool kit for a craftsperson.

The following is a list of tools used for the application of wallcoverings:

- paperhanging brush/sweep
- trimming knives
- tape measure (3–5 m)
- chalk line
- felt roller
- straight edge
- casing wheel
- buckets and troughs
- paste table
- shears or scissors
- folding rulers
- plumb bob and twine
- boxwood roller/edge roller
- spirit level
- ridgley trimmer
- paste brush
- hand trimmer (morgan lee) for papers with selvedge
- caulker.

Key point

When painting and decorating it may be necessary to carry out minor repairs such as:

- Removing and replacing a broken or cracked pane of glass in a window
- Removing and replacing a damaged section of skirting board
- Fixing Dado Rail or a picture rails to wall surfaces at the customers request
- Repairing defective plaster areas after stripping wallpaper.

The following are items of equipment used in painting and decorating tasks:

- propane gas torches
- steam strippers
- orbital sanders
- pneumatic sanders
- troughs for applying Vinyl wallpapers
- extension cable
- paste table
- hot air guns
- palm sanders
- rotary wire brushes
- pasting machines
- vacuum.

The following are useful objects and items that help with painting and decorating tasks:

- pot hook
- sponge
- gauzes/strainers
- kettles
- rubbing block
- trays
- brush keeps
- window scrapers
- pan and hand brush
- masking tape
- dust sheets
- knotting pots
- sanding pads
- tac rags
- roller sleeve cleaner
- speed strippers.

1.2
Hand tools

1.2.1 Description and use

Paint scraper or broad knife
Sizes are available in blade widths of 25/50/75/100 mm. A good quality knife has a hardwood handle riveted to tempered steel blade, but some are manufactured with polypropylene handles.

These knives are used to remove loose and flaking paint, wallpapers and fabrics, bits and nibs.

Figure 1.1 *Paint scraper or broad knife*

Figure 1.2 *Filling knife*

Figure 1.3 *Chisel knife*

Figure 1.4 *Putty or stopping knife*

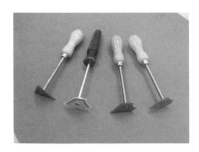

Figure 1.5 *Shave hooks*

Filling knife

Sizes are available in blade widths of 25/50/75/100/150 mm. After use clean the blade and protect the edge with a plastic cover strip.

The filling knife is similar in construction to a scraper but the blade is more flexible to allow working of the filler into undulations in surfaces.

The knife is used to fill open grain work on timber shallows and indentations, cracks and holes.

Chisel knife

Blade sizes are available in widths of 25–50 mm. These knives are constructed in the same way as the scraper. The chisel knife is used to remove paint, wallpaper, bits and nibs where the scraper does not fit. On mouldings, between door frames where the broad knife will not fit and on window and door mouldings.

Putty or stopping knife

Used for applying facing and bedding putty to traditional wood and metal windows when glazing or replacing broken or cracked panes of glass. The knife is also used to fill small nail holes, open timber joints and cracks on woodwork.

Shave hooks

A tool consisting of a hardwood or polypropylene handle. On the hardwood handle type, a circular metal spindle goes through to the scraping head. The scraping head is bevelled and must be kept sharp (use a file). There are three scraping heads that can be obtained pear shaped, triangular and the combination. The shave hook is used to remove paint and varnish coatings from mouldings and beadings when used with heat guns or paint strippers.

Palette knives

These knives come in many shapes and sizes and there are two main types:

- For artists with flexible thin blades as illustrated
- For painters with flexible blades.

The main use of the knives is to mix pigments/paints together to colour match to other materials such as fabrics, or to customer's requests. Once the samples match, larger quantities can be mixed and produced to carry out the painting task.

Figure 1.6 *Palette knives*

1.3
Basic tools

Paint stirrers/agitators
Stiff metal blade housed in a hardwood or plastic handle. The blade has holes placed along its length to enable the oil and pigment to disperse and come together as a mixture. The stirrer is used to agitate settled paint back to its correct liquid thickness before application by brush or roller.

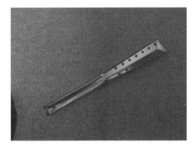

Figure 1.7 *Paint stirrers/agitators*

1.3.1 Description and use

Painters and decorators should through time collect a selection of basic hand tools to enable minor repairs to be carried out during the process of preparation for painting and decorating. This will enable the applied decoration to be carried out to a satisfactory conclusion and enable a higher standard of finish.

Hammers
These are necessary to carry out a number of tasks when preparing surfaces, such as the removal of protruding nails from timber skirting, architrave and rails. When used with nail punches to drive nails and pins used for fixing dado rails below the surface, filling or stopping can then be carried out thus hiding the fixing hole. Pincers and pliers can be used for many tasks including the removal of nails, wire clips

Figure 1.8 *Ball*

Figure 1.9 *Claw*

Figure 1.10 *Pin*

Figure 1.11 *Pincers*

Figure 1.12 *Pliers*

Figure 1.13 *Caulking tool*

holding telephone cables to the tops of skirting, the trimming of wires if rewiring a 13-amp electric plug and extracting picture hooks from wall surfaces.

The caulking tool can be used for stopping holes and undulations on plastered surfaces to achieve a flatter surface to apply decoration.

Wire brushes can be used to clean metal surfaces prior to painting, removing loose material from brickwork or for scouring the surface of wallpapers to enable speedier soaking prior to removal.

Chisels would be used to remove beadings from windows that require renewing or for effecting minor repairs to timber.

Screwdrivers are used for the removal of fixtures and fittings.

Figure 1.14 *Wire Brushes*

Figure 1.15 *Dust pan and brush*

Figure 1.16 *Nail punch*

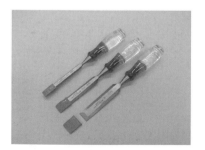

Figure 1.17 *Chisel selection*

Figure 1.18 *Screwdrivers*

Figure 1.19 *Head detail*

Figure 1.20 *Pointing trowel*   **Figure 1.21** *Dusting brush*

Figure 1.22 *Pots and kettles*

1.4 Paperhanging tool kit

1.4.1 Description and use

Paperhanging brush/sweep
These brushes are manufactured with a hardwood handle of Ash Alder
or Beech to give a comfortable grip and not cause blisters to the operatives hand, the filling is pure bristle. The filling is set in a synthetic
rubber setting of vulcanized rubber to prevent loss of bristles. Cheap
brushes have synthetic fillings of nylon but these have no body and do
not remove all air from behind the paper when smoothing out during
the application process. Sizes vary from 150/200/250 mm.

When hanging a covering such as a fabric or wallpaper the brush is
used to remove all air bubbles between the coverings and the wall.
The soft bristles do not scratch/indent or mark the covering.

Do not use a sweep when hanging traditional flock papers. Caulking
blades can be used as an alternative to a sweep when hanging vinyl
papers or fabrics, but only on flat, even walls and ceilings.

Trimming knives
There are many manufactured types and have either fixed, retractable
or snap off blades.

A selection of all types is a must for the paperhanger's toolkit but
be sure to purchase good-quality products that can meet the function
of use.

Figure 1.23 *Paperhanging
brush/sweep*

Figure 1.24 *Trimming knives*

Figure 1.25 *Rulers, measures and tapes*

Keep a selection of replacement blades at all times and have available a container which can house blunt blades.

New and replacement blades are provided in disposable cartridge cases to prevent injury to self or others.

Trimming knives are usually named after the manufacturer, for example, Stanley Knife. These have either fixed or retractable blades. To change the blades it is necessary to unscrew the housing and locate the blade on the positioning lugs. Spare blades can be kept inside the housing.

Xacto knife is a small, round slim-handled stem with a scalpel-type blade inserted into a holding device. This device is turned clockwise to secure the blade and anti-clockwise to remove it. Snap off knives are available which are economical to buy with sleeves of replacement blades, however these are flimsy and the blades when blunt are not always easy to break or snap off.

Rulers, measures and tapes

Measures are constructed from one-metre strips of hardwood with markings in metric on one side of the measure and imperial (feet and inches) on the other. Measuring rods are also available.

The decision to use a measure rule or tape depends upon the type of task to be undertaken.

Panel work
Application of Decorative
Features
Mixed Papers
Borders and Mitres

Tapes are flexible metal strips of various lengths housed in metal or plastic casings. They are ideal for measuring surface areas but limited in use when measuring and cutting coverings to accurate tolerances on the paste table. The tapes have a locking device on the casing to hold the tape open at any measurement setting. When not in use the tape retracts into the casing for protection from damage.

Rulers are usually one metre in length, constructed from four sections of boxwood or PVC. They fold into sections when not in use for ease of storage. One side of the ruler shows metric markings, the other side imperial.

When using measuring implements during paperhanging keep them clean and free from paste, do not use them as substitute-straight edges when trimming coverings as they will certainly be damaged by trimming knife blades, and never flip the sections up and down as this action can damage the hinges holding the sections together.

Look after your tool kit and it will last you a working life.

Figure 1.26 *Plumb bob and line*

Figure 1.27 *Spirit level*

Figure 1.28 *Straight edges*

> **Key point**
>
> When trimming paper on a paste table use a protective zinc strip under the paper or fabric to prevent damage to the table. A sharper trim cut will also be obtained.

Figure 1.29 *Rubber rollers*

Plumb bob and line

A weighted pendulum to which a twine line is attached.

To obtain the vertical guide line suspend the plumb bob from the string. When it is still mark inverted 'v's behind the twine at intervals of 300 mm down the height of the wall. The edge of the paper is hung to these guide marks. An alternative method is to chalk the string, press the line to the wall and snap (twang) the line. This will leave a chalk guide line.

Spirit levels and straightedges

Spirit levels are used by painters and decorators to apply vertical and horizontal lines to surfaces and to check that applied decoration is vertical or horizontal. The inserts are glass tubes filled with fluid and containing an air bubble. The tube has location marks on the outside of the glass tube. When the level is held directly vertical or horizontal the bubble is in the centre of the location marks indicating accuracy.

This device allows the paperhanger to mark out accurately on surfaces, division areas such as friezes/dado's/wall fillings/paneling and borders.

Straight edges are bevel edged lengths of steel or wood used for a variety of activities. They have to be perfectly straight with at least one edge along its length bevelled. They come in various lengths from 300 mm to 1 m. They are used with trimming knives to trim applied decoration by producing butt joins, mitres and shapes on the surface when hanging wall coverings.

Rollers – Rubber/felt/seam

Rubber rollers consist of composite hard solid rubber set in a Roller arm. It is used as an alternative to a paperhanger's sweep to smooth out heavy vinyl's or murals during application to walls where a sweep would not be heavy enough to roll out the air pockets. Sizes are in the range of 90–175 mm.

Felt rollers are used instead of a sweep to apply delicate wall coverings. They are constructed of a number of small felt discs which are housed on a central spindle. When used correctly the roller applies even pressure without polishing and damaging the surfaces of wall coverings. Wash and dry after use. Sizes are in the range of 90–175 mm.

Seam roller's heads are made from materials such as boxwood or plastic and have single or double arm frames. Their purpose is to roll down the edges of applied wall coverings. At joins, angles or seams the roller face must be kept clean at all times and must never be used on embossed papers. The roller will flatten the relief pattern leaving flattened visible stripes at the joins. Sizes are various.

Shears or scissors

The best quality traditional paperhanger's shears are constructed from polished stainless steel blades, which are hollow ground. The hollow grinding allows space between the blades when cutting pasted paper,

Figure 1.30 *Felt roller*

Figure 1.31 *Seam roller's*

Figure 1.32 *Shears or scissors*

Figure 1.33 *Ridgley trimmer and straightedge*

Key point

Care must be taken when using rollers during the paperhanging process. If too much pressure is applied the emboss on some wallpapers could be compressed leaving unsightly marks.

---

thus preventing tearing and giving a good sharp cut. The more modern and perhaps more frequently renewable type of shears are flat bladed, more lightweight in feel and the tips of the blade are flat, which can be deemed a disadvantage when trimming the waste. Sizes are from 250 to 300 mm long in length.

Care and use

Never use paperhanger's shears for cutting other products for they will become blunt. Keep the blades clean and free from adhesive during and after use by frequently washing in warm soapy water.

Never use abrasive paper to sharpen the edge and do not over tighten the tension screw. If the shears are kept clean it is very rare that sharpening of the blades becomes necessary. If sharpening is required use an oil stone, hold the blade edge at a 45° angle to the stone and make a single swift pass along the stone. Repeat this several times and the blade would be sharp again. Use machine oil as a lubricant.

Ridgley trimmer and straightedge

A cutting wheel mounted in a guide with a spring-loaded handle which when pressed down engages the cutting wheel with the paper to be trimmed, when released the blade lifts away from the paper face, thus giving a precision cut. This assembly slides on to a locating slot on a machined stainless steel straightedge. A zinc metal strip is located in slots under the straightedge.

To use the equipment remove the zinc strip and place it on the paste table. Locate the wall covering where the cut is to occur, directly over the zinc strip. Place the stainless steel straightedge on top of the wall hanging and locate the cutting wheel in the slot. Using the palm of the hand press down on the handle to operate the cutting wheel and push along the straightedge. This will accurately trim the wall covering to the desired dimensions.

Pasting brush

An ideal brush to use when applying adhesive to wall coverings should be a worn-down flat wall brush. With the bristles being worn down to approximately half their true length the brush distributes the adhesive more evenly. These brushes should be selected according to the user from 100 to 175 mm. These brushes are made from a mixture of pure bristle and horse hair. The ferrule should be made from copper. This eliminates the rusting of the ferrule when using water based adhesives. Always wash paste brushes out in soap and water, dry and store after use.

Paste table

These are folding leaf tables with supporting legs that open up from beneath the table and lock into place. They are lightweight in

Activity 1.1

Produce a list of equipment required to carry out the following tasks

- remove wallpaper from walls
- cut and trim wallpaper on a paste bench
- apply wallpaper to walls.

Figure 1.34 *Pasting brush*

construction for ease of transport and contain a carrying handle. The following activities are carried out by the paperhanger on the paste table:

- checking of all the rolls before unwrapping to determine the batch and shade numbers are the same
- that the paper is not damaged
- measuring of required lengths
- matching of lengths
- cutting and trimming of lengths
- pasting and folding.

Table size is 1.830-m long by 0.560-m wide when in the open position.

The table tops are constructed from hard pressed hardboard or marine plywood for durability; some tables are available with inbuilt measuring devices.

Note that for table maintenance keep table tops and edges free from adhesive because dried adhesive can damage new wallpaper. Check all catches, handles and hinges periodically.

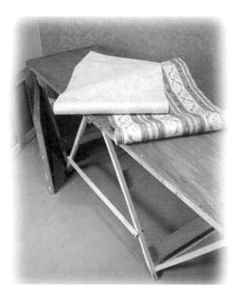

Figure 1.35 *Paste table*

Test your knowledge 1.2

1. Select from the illustrations shown and name the items that would be used to check that wallpapers have been hung vertically or horizontally.

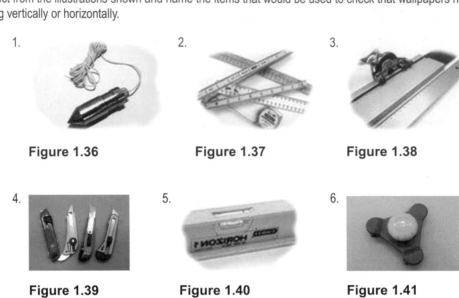

1.	2.	3.
Figure 1.36	**Figure 1.37**	**Figure 1.38**

4.	5.	6.
Figure 1.39	**Figure 1.40**	**Figure 1.41**

2. What is illustration 6 used for?
3. What is illustration 5 used for?
4. For which of the above, rules and measures should not be used as makeshift?

1.5 Equipment used in painting and decorating tasks

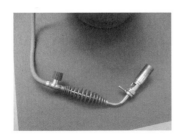

Figure 1.42 *Propane gas torches*

1.5.1 Descriptions/purpose

Propane gas torches

These items of equipment are used with a bottled supply of propane or butane – an economical fuel that allows the removal of coatings within an acceptable timeframe. They are used by painters and decorators to remove paint and varnish coatings by a heat process. The gun or torch is attached to a hose and regulator. This equipment has replaced the traditional painter's blowlamp which was fuelled by petrol or paraffin.

Hot air gun

An electrically operated gun which is used to remove paint and varnish by heat. An element is heated and air is drawn through it. This heated air is directed on to the surface and softens the coating, which can then be removed by using shave hooks and scrapers. The disadvantage of this piece of equipment is that when used outside the heat is dispersed before the paint coating has softened due to wind and air influence.

Figure 1.43 *Hot air gun*

Figure 1.44 *Steam stripper*

Figure 1.45 *Palm sander*

Steam stripper

The steam stripper is a huge kettle. It boils up water, turns it into steam and this steam is directed to the surface of wallpaper to soften it prior to removal. The equipment consists of a water vessel and filler gauge, a filler cap, steam hoses and perforator plates. When the water boils and turns into steam the hose directs this steam to the perforator plate which is held against wallpaper. The wallpaper is softened by the steam and can then be removed from the wall surface using a scraper.

Palm sander

An electrically powered lightweight sander for preparation of surfaces prior to the application of surface coatings and between coatings. Where possible use a transformer with a 110-volt electricity supply for safety reasons.

Orbital sander

An electric sanding pad that rotates on a cam attachment. The sanding pads can be bought to fit or can be cut from standard sheets. Orbital sanders can be obtained to operate with electricity or compressed air.

Rotary pneumatic sander

An air-operated sanding appliance used with water and compressed air. The compressed air required to power pneumatic sanders is 80 psi or 5.5 bar (15 psi = 1 bar).

Orbital pneumatic sander

These sanding tools are powered by compressed air and used with a lubricant such as water.

Pasting machine

Wallpaper is inserted between rollers and is drawn through a paste trough housed under the rollers. The backing of the wallpaper is coated evenly with adhesive and the paper can be folded or applied directly to wall surfaces depending on soaking time of the product. Standard dimension rolls can be used in this machine.

Figure 1.46 *Orbital sander*

Figure 1.47 *Rotary pneumatic sander*

Figure 1.48 *Orbital pneumatic sander*

14   Tools and equipment

Key point

Use personal protective equipment (PPE) such as goggles and gloves when using pneumatic sanders. As with all power tools either electrically or pneumatically operated, training should be given before the worker uses such equipment.

Extension cables
Electric cable providing 240/110-volt supply to power tools. Many lengths are available but before use the cable must be fully unwound to prevent heating up of the cable core. The recommended maximum length of cable should not exceed 20 m.

110-volt Transformer
This item reduces 240-volt supply to 110 volts. Electrically powered tools operating off this voltage are much safer for the operator.

Vacuum cleaner
In the painting and decorating profession the use of a vacuum cleaner at the end of a job or days' work does wonders for the reputation of the worker in the eyes of the private customer. Industrial cleaners would be used on site to clean up dust and debris.

Brush keep
This piece of equipment is used to store brushes in the wet state, by excluding air from the container. A solvent is placed in a bottle with an evaporation wick. The fumes from the evaporating solvent replace the air in the container thus preventing the paint on the brushes drying out. It saves money, labour and resources that would normally be used for cleaning brushes after use.

Figure 1.49 *Pasting machine*

Figure 1.50 *Extension cables*

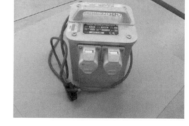

Figure 1.51 *110-volt Transformer*

Figure 1.52 *Vacuum cleaner*

Figure 1.53 *Brush keep*

Key point

- Purchasing cheap tools and equipment is false economy
- Use the correct tools and equipment for the job to be undertaken
- Clean all tools and equipment after use
- Read manufacturers instructions prior to the use of tools and equipment
- Protect flexible blades with protective strips
- Use the correct set up when using 240-volt or 110-volt supply
- Never use faulty equipment
- Ensure that electrical equipment has a 'test for use' sticker on the appliance
- Fully unwind extension cables prior to use
- Never trail electrical cable through water
- Wear appropriate PPE whilst using tools and equipment.

Key point

Safe use of tools and equipment

Before using any item to carry out a task of work you should seek instruction on its use. It is your employers duty to train you. It is your responsibility to work safely to ensure you do not endanger yourself or others. Always read instructions on use of/and maintenance of all hand tools. Seek instruction if you are not familiar with an item of equipment. Clean all tools and equipment after use, they will last longer and perform the tasks which are required of them. Make regular tests on all electrical/pneumatic tools. Do not use any item of equipment that appears damaged or its user date for checks has expired.

Work safe!

Test your knowledge 1.3

Place a, b, c or d in the box alongside the question.

1. The hand tool used to remove paint coatings from timber mouldings after they have been heated up is a?
   a) scraper
   b) putty knife
   c) palette knife
   d) shave hook
2. The item of equipment used to check that wallpaper borders run horizontal after hanging is?
   a) plumb-bob
   b) straight edge
   c) spirit level
   d) ruler
3. What item of equipment reduces 240-v electrical supplies to 110 v for safety reasons?
   a) vacuum
   b) extension cable
   c) transformer
   d) hot air gun

2 Preparation of surfaces

2.1 Preparation of previously painted surfaces

There are two factors that will decide the type of preparation required prior to the application of decoration

- If surfaces are in good condition and have been previously painted, the only action that will be required prior to decoration will be to wash down, make good and apply a fresh paint system.
- If surfaces have started to show signs of film breakdown, it will be necessary to completely strip the surface. Once stripped surfaces will require a three coat paint system, but for best quality work specify a four coat paint system.

The breakdown of paint coatings can be caused by:

- wear and tear of everyday environmental traffic, both internal and external
- poor initial preparation
- using cheap products
- vandalism
- industrial, coastal or rural atmospheres.

Wear and tear – The use of the environmental space by the population both at work and in leisure. Light degradation, physical abrasion and weathering are contributing factors.

Poor initial preparation – If preparation processes prevent coatings gaining a key (grip) either by specific or mechanical adhesion. The final result will be early breakdown of the system. It could be due to lack of sanding the surface, not cleaning off paint residue after sanding, applying coatings and trapping moisture, incorrect filling plus many other reasons.

Activity 2.1

Carry out a group discussion and list the consequences of careless preparation in relation to customer relationships. Hand in your list for inclusion to your portfolio of evidence.

Use of cheap alternative products – It is false economy to select and use cheaper paints. These cheaper products do not stand up to the continuous environmental conditions. Cheaper alternatives are usually bulked up with extenders and poor quality resins which offer little resistance to the natural elements.

Vandalism – Graffiti caused by vandalism has to be removed due to the unsightly nature of the applied product. This often results in redecoration or cleaning of the spoiled surface.

Industrial environments – Emissions from industrial plants can result in acidic products being deposited on exterior surfaces by the natural elements. This acidic product when it settles on surfaces attacks paint coatings and breaks them down resulting in the need for redecoration.

Coastal environments – Sea frets, fog, mist and low cloud in coastal areas can deposit alkali on to surfaces. This alkali can quickly break down paint systems.

Rural environments – The chemicals used on the land by farmers can be transferred to surfaces by wind causing damage to paint systems by abrasion.

Expansion and contraction – Continuous temperature changes of surfaces eventually causes breakdown of paint systems.

Washing down – Detergents can be used when washing down a surface in good condition. These products will remove grease, grime, dirt and nicotine. The surface should then be rinsed with clean water and left to dry before applying any decoration.

Previously painted surfaces – Surfaces that are thoroughly prepared will enable the applied paint film to adhere and bond to a surface. The life of the paint system will last up to four times longer. The surface itself will then have a longer life span due to the protective nature of the applied paint system.

The following are the methods of preparation for previously painted surfaces prior to redecoration:

- washing down using sugar soap or detergent
- minor repairs to fine cracks and indentations with proprietary fillers and stoppers
- abrading by hand with wet or dry processes

Key point

It must be understood that water can soak through some paint coatings into the surface. This causes that surface to swell; they shrink back on drying out. This causes great strains on the adhesive qualities of any paint film that has been previously applied to a surface.

Key point

Many of the above preparatory methods will be described in greater detail in other sections of this chapter.

Test your knowledge 2.1

1. If a previously painted surface is in good condition what preparation should be carried out prior to repainting?
2. Name the items of equipment that can be used to aid the preparation process.
3. Name the contributory factors that result in the breakdown of paint systems.
4. How do climatic weather conditions breakdown paint systems? What is deposited on surfaces?

- dry abrading using electrical appliances
- wet abrading using mechanical (pneumatic) appliances
- removal of coatings using heat
- removal of coatings using chemicals
- major repairs to surfaces due to damage
- removal of surface defects caused by previous poor application.

Activity 2.2

Write a short report on temperature change in relation to expansion and contraction of building surfaces and how it can affect applied coatings. Hand in your report for inclusion as portfolio evidence.

2.2 Preparation of timber surfaces

The painter and decorator needs to

- identify by recognition softwoods and hardwoods
- have some knowledge of their composition and nature
- be able to specify the correct preparatory treatment.

Timbers are classified by the term 'softwood' or 'hardwood' and terms used to enable recognition are as follows:

- coniferous – evergreen – needle shaped leaves – quick growing – resinous
- deciduous – seasonal – sheds leaves in autumn – slow growth

Coniferous trees are softwoods and include pines, cedars and spruces. Deciduous trees are hardwoods and include oak, beech, mahogany and walnut.

Softwoods are used for first and second fix joinery work, components such as skirting, dado and picture rails, architraves, door and window casings, barge boards, fascias and soffits and some claddings. Hardwoods are more expensive to buy and are used for exterior doors and windows or where a more visually stimulating timber effect is desired such as panelling or veneer inlay work. Softwoods are painted due to the knotty and resinous nature of the timber. Hardwoods require a finish in clear coatings using varnish, lacquer or French polish to enhance the beauty of the grain.

Activity 2.3

Compile a list of both hardwoods and softwoods. From the Internet cut and paste pictures of the trees to match the tree type. Submit your presentation for inclusion to your portfolio.

(a)   (b)   (c)

Figure 2.1 *Spruce trees/leaves/notice the needle like leaves*

Timbers are porous materials by nature:

- Good key for clear and pigmented coatings can be obtained.
- They absorb water easily therefore must be protected.

Moisture in timber
- *Absorption of water* – The cellular structure of wood is capable of holding large quantities of water; if timber has been properly seasoned it should contain no excess moisture. Contact with water or moisture must be prevented before any protective coating is applied.
- *Moisture content* – Timbers used internally should have 10–20% moisture content. Rising to 15–18% for exterior woodwork.

If timber has high moisture levels it is likely that applied coatings will breakdown through a series of defects such as

- blistering caused by seeping resin or moisture
- wet rot leading to dry rot
- infestation of mould and fungi.

The moisture content of timbers prior to the application of any coating on new and previously painted timber should be measured with a meter if it is suspected that the timber is above acceptable levels.

Seasoned timber   this is wood that has been treated to reduce its moisture content to make it stable

- In natural seasoning the wood is gradually dried out in the open air. It is a lengthy process and for this reason most timbers are artificially force dried.
- In kiln dried seasoning the moisture content is much lower than naturally seasoned timber. When this timber is used to construct components such as windows and doors, the components absorb moisture after they are fitted to the building structure; this causes problems for the painter through expansion of the door or window resulting in joins opening up.

Softwoods

Softwoods should be lightly abraded with fine glass paper, diagonally across the grain to scuff the surface finally finishing off very lightly in the direction of the grain. Abrading enables the mechanical adhesion process of the applied coating. Never sand across the grain as the sanding marks show through the applied coating and look unsightly.

New or bare wood preparation is as follows:

- Select and use F2 glass paper, sanding with the grain
- Punch any nails or pins below the surface and seal to prevent staining or corrosion
- Remove surface dust and apply two coats of shellac knotting to all resin ducts allowing 10 minutes between coats
- Remove loose knots and replace with wood plugs

The timber should now be ready to receive its first coat of the specified paint system.

> ### Key point
>
> Internal priming of external doors and windows should always be carried out using exterior quality primers.

Activity 2.4

On new or uncoated internal timber surfaces, quick drying acrylic water based products are preferred, they have no strong fumes. Acrylic systems are available in primers, undercoats and glosses and can be applied in one day.

Explain through discussion with your group how this could be beneficial to the customer. List the advantages. Submit your evidence for inclusion in your portfolio.

Hardwoods

The cellular structure of hardwoods is a lot finer than that of softwoods and this causes an adhesion problem for clear coating systems especially on oak and ash. Some of the hardwoods have an oily nature, others are acidic. Aluminium primers with leaf or flake pigment enable good specific adhesion.

Surface preparation of hardwood is as follows:

- Sand with the grain, use the correct grade of abrasive
- Dust off the surface and degrease with turpentine
- Dry off the surface and apply the coating system.

End grain

End grain of timber is more absorbent than any other face of the wood. The cells conduct moisture up the capillary tubes through which moisture will travel. It is essential when painting timber that all end grains should be properly primed to prevent this absorption. A sound practice is to apply two coats of shellac knotting prior to the application of the primer.

Figure 2.2 *End grain*

Figure 2.3 *Knotting pot*

Test your knowledge 2.2

1. Timbers are classified into two groups. Name the two groups?
2. Which types are usually painted and which types stained and varnished?
3. What natural defect is present in softwoods and how can it affect applied paint systems?
4. What substance is applied to this natural defect to form a barrier between its exudations and any applied paint coating?
5. When abrading timber surfaces which direction should be followed to ensure that the surface is not damaged?
6. Name some softwoods and hardwoods?

Resin and knots in timber

If the exposed surface of the knots is not sealed before paint is applied there is a danger of bleeding of the resin into the paint film causing discolouration and staining. Knots that exude copious amounts of resin should be drilled out and plugged with wood. Normal treatment, apply two coats of shellac knotting allowing 10 minutes between coats. Ensure the coats are thin and well brushed out with no edge build up. Apply beyond the area of the knot or resin seepage.

Decayed and denatured wood

When wood is decaying due to wood rot it should be cut out completely and replaced with sound timber. Sometimes it may be necessary to remove the whole component and replace with a new component. The new insert and the remainder of the old wood should be treated with a fungicidal solution before repainting denatured wood. When wood is exposed to the weather for a long period of time the cellular structure of the wood begins to breakdown, leaving a surface which becomes furry or fibrous. It is extremely difficult to get a paint to adhere to such a surface.

Preparation of denatured timber

- Abrade the surface with a fine grade abrasive to remove the dead fibres and dust off
- Treat the surface with a wood preservative and allow to dry out
- Apply a coat of raw linseed oil to the surface and leave for 15 minutes
- Remove the surplus oil with a cloth and leave for a week to dry out so the remaining absorbed oil can oxidise
- Repaint with selected paint system.

Building boards

The painter is required to apply paint systems to many surfaces to make them visibly acceptable. Building boards are included.

The following are the types of building board:

- Medium density fibreboard (MDF)
- block board
- plywood
- dry lining board/plasterboard
- hardboard.

Boards need to be prepared according to the nature of the product to ensure the product is not damaged. Things to consider are:

- composition of the board/absorbency
- Are there health risks or hazards related to preparing such surfaces
- What type of abrasive should be used
- What type of primer or sealer should be specified for use
- What coating system should be specified
- preferred application method.

Activity 2.5

Search the Internet and obtain pictures of each of the building boards. Read paint specifier manuals and recommend an appropriate primer or sealer for each of your selected boards.

If resources permit obtain samples of board approximately 150-mm square and apply primer sealer to half of the sample. Make brief notes on sinkage/absorbancy.

2.3
Preparation of metal surfaces

To assess metal surfaces prior to preparation and painting we should consider the following:

- are they ferrous or non-ferrous metals
- are we familiar with the causes of corrosion and its mechanism
- mill scale and rust are different products
- how the environment affects metal surfaces
- can the correct preparation methods for different types of metals be specified
- can we recognise, identify and select tools and materials to prepare metals
- are we able to specify paint systems for use on various types of metals.

The following list identifies how metals can be damaged by environmental effects

- acids and alkalis
- moisture
- grease, dirt and excessive manual handling
- temperature change.

The following factors for the successful preparation and painting of metals should be considered

- Correct surface preparation to a standard (Swedish standards)
- A suitable paint system for ferrous or non-ferrous metals should be specified
- The painting procedure including the application of the system at the appropriate time/under good conditions with adequate film thickness.

A pre-check of surfaces should be carried out before any paint is applied to check for traces of the following:

- mill scale or rust
- oil/grease/wax/dirt

- loose old paintwork
- weld/scale/flux
- acid/alkali attack
- moisture/exposure.

2.3.1 Preparation methods of metals

Rust and corrosion can be present in various degrees; this will determine the method of preparation required. For the removal of light rusting use the following hand preparation methods:

- abrasive papers and cloths
- manual scraping
- manual wire brushing
- derusting jellies or phosphate solutions
- degreasing solvents.

If corrosion is further advanced, use the following methods of preparation:

- shot/sand or grit blasting
- flame cleaning
- needle guns
- mechanical chisels/wire brushes/wheels
- pickling (acids).

All the above require training of personnel as they are specialist methods of preparation.

Metal is prepared in any location, usually *in situ* on site. Some of these locations could be highly dangerous environments such as chemical works. In this case bronze–phosphor tools have to be used to prevent the risk of sparking which could lead to explosions of gases or chemicals present in the local atmosphere.

Corrosion is caused by high relative humidity and atmospheric pollution. The following is an example of the corrosion cycle

- Salts which can be present in the atmosphere lands on the metal surface
- These salts are deliquescent in nature and absorb moisture or water readily thus accelerating the corrosion and rusting process
- Ferrous sulphate (sulphur) in the atmosphere is deposited on to the steel by rain and becomes concentrated at angles and arises
- This leads to the formation of high rates of rust product which then falls off in sheets as it weathers
- Scraping and wire brushing is an inadequate method of preparation to remove ferrous sulphate (which eats away the metal).

2.3.2 Methods of removing mill scale, corrosion or rust

- *Weathering* – Corrosive products can be loosened by the action of atmospheric conditions such as expansion and contraction caused by temperature change upon the metal.
- *Shot blasting* – A system of cleaning corroded metal by projecting at high velocity, sand, grit or shot onto the surface. Any corrosive product present is loosened from the parent metal leaving it clean and ready for repainting or treatment.
- *Flame cleaning* – Hot oxy acetylene flame is applied to the surface, this causes the rust to expand at different rates to the metal. Eventually through expansion and contraction the rust can be easily scraped off. Do not use on thin gauge metals.
- *Mechanical needle guns* – A series of steel needles are vibrated on corroding metals using compressed air to remove contamination. Damage can be caused if this method of preparation is used on thin gauge metals.

2.3.3 The mechanism of corrosion

If two different metals are present in an electrolyte which could be rainwater one of the metals will become sacrificial to the other. An electrical current can be detected by the use of a voltmeter. The presence of this electrical current causes one of the metals to corrode.

This electric current exists when there are different metals present and they have a circuit completed by the presence of water. Remove the moisture or place a barrier between the metals and the circuit is broken.

The three elements that contribute to causing corrosion are:

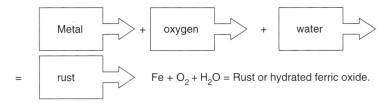

Figure 2.4 *The three elements for rusting to occur*

Water comes into contact with metals as

- rain, polluted as it passes through the atmosphere
- salt spray in coastal areas
- condensation that is formed during humid conditions
- dampness given off from soils and building materials

Metals should never be painted if the temperature is below 40° Fahrenheit or the relative humidity of the air is above 80%.

Test your knowledge 2.3

1. Which of the following is a ferrous metal?
   - Aluminium
   - Copper
   - Steel
   - Lead.

2. The following metals corrode, but which one rusts?
   - Copper
   - Zinc
   - Aluminium
   - Iron.

3. Metals should not be painted when the relative humidity of the atmosphere is above
   - 15%
   - 30%
   - 80%
   - 60%.

4. The minimum recommended thickness of a paint system on metal should not be less than
   - 25 microns
   - 50 microns
   - 75 microns
   - 100 microns.

5. The preparation of ferrous metal using temperature to loosen and remove rust is termed
   - blast cleaning
   - flame cleaning
   - pickling
   - weathering.

Hidden problems under the paint film   Moisture can permeate through some paint films allowing ferric oxide deposits to form corrosive products such as rust under the coating. The deterioration of the metal can continue for long periods of time until the structural integrity of the metal is compromised.

The thickness of a normal coat of paint is 25 microns, on metals the minimum recommended thickness of the paint system is 75 microns.

Mill scale is produced by the rapid oxidisation of white hot metal when it comes into contact with cool air as it leaves the furnace. It is the skin formed on the surface of the molten metal. If it is not fractured in any way, it protects the parent metal, but weathering usually loosens it through expansion and contraction.

The painter needs to apply the protective paint system to newly erected structures as soon as possible to prevent the corrosion process.

Figure 2.5 *Steel structures*

2.4
Preparation of plaster surfaces

Plasters, brickwork, block work and concrete surfaces are all absorbent by nature and require the application of protective to make the surface look presentable.

To enable paint systems application it will be necessary to have knowledge of the following:

- the types of plaster used in construction from previous and present practices
- how the nature of these surfaces affected specified coatings by physical and chemical means

- Defects that can be encountered due to attack by salts and moisture.
- Be able to specify the remedial preparatory work required to prevent such defects and specify paint systems that will not be affected by the return of salts and moisture.

2.4.1 Decorative plasterwork

There are many skills involved with the application of paints, stains and varnishes as there are sectors within painting occupations. The conservation decorator would be involved with the restoration or maintenance of decorative plasterwork or ornamentation. It requires special skills to carry out work such as:

- Colour mixing and blending to match faded previously applied paints
- Gilding of plaster, wood or glass
- Sign writing or lettering such as manuscript work
- Template making to repair enrichments
- Accuracy of applied decoration.

Figure 2.6 *Decorative plaster cornice*

One of the major contributing factors of historic buildings becoming damaged is the presence of damp as most of the old construction materials had a large presence of salt in the plaster composition. Moisture and damp reactivates these dormant salts causing the ornamentation to loose its decoration through spalling.

Moisture content of plaster before any surface coatings are applied to plaster surfaces ensures that

- the plaster is dry and no moisture remains
- there are no damp patches
- defects are made good
- carry out the correct preparation and apply the specified paint system.

Key point

Use a moisture meter and universal indicator to check for alkali or moisture presence in surfaces. Universal indicator if placed on a damp wall that contains alkali will turn blue. Litmus paper could be used as an alternative.

Saponification – A defect caused by free alkaline deposits combining with the oil or resin in oil based paints. The alkali salt turns some oils and resins into a non-drying sticky brown solution. The paint is discoloured and cannot dry. Oil based paints can be safely applied to such surfaces if no moisture is present but any penetration of water at a later date could reactivate any salts that are present which will then attack the dried film.

Dry-out – A defect caused by the rapid evaporation of water from the applied skimming coat or top coat of finishing plaster. If the browning or backing plaster is too porous or artificial heat is used to speed up the drying process the top coat can flake off especially when rollers are used to apply the surface coating.

Efflorescence and associates – Efflorescent salts are alkaline in nature and are present in sand used during the building construction stage. When the surfaces start to dry out the moisture in the surface evaporates into the atmosphere and the salt is left on the surface of the building material and appears as a white deposit.

Deliquescent salts – These are hygroscopic salts that absorb moisture from the air into the surface. They appear as damp patches on plaster surfaces.

Crypto-crystalline salts – Are salts that absorb moisture and grow into crystals which in turn push off the plaster damaging the plaster finish?

2.4.2 Process of preparation if salts are present on the surface

Dry brush the salt from the surface, do not wash off with a sponge and water as the salts will turn back into solution and re-enter the surface. If there is no further reaction after observing the surface for a few days, it can be considered safe to repaint. On porous surfaces which may be suspected as having alkali content, it is usually considered best practice to apply water based paints such as emulsions, new design distempers or acrylics. The structure of these coatings allow the movement of moisture without damaging the coating, thus allowing any salts present to escape through the coating to lie on the surface and be brushed off at a later date.

If the specified paint system has to be oil or resin based use special primers that are not attacked by alkali. The medium of these special paints contain tung oil, phenolic or coumarone resin, these are not affected by alkali.

Alkali resistant primers or chlorinated rubber primers are the best coatings; they form a barrier and allow the conventional oil paint system to be applied.

Test your knowledge 2.4

1. Plasters and brickwork are porous or absorbent surfaces. Explain how this can affect the application of water based paints.
2. Why are new plaster surfaces never sanded with coarse abrasive prior to the application of the sealer or primer?
3. What can be used to test for alkali deposits in new plaster walls?
4. Describe how the presence of salts in brickwork and plaster can affect the application of paints?

Add your responses to your portfolio of evidence.

Preparation of new plaster

- De nib the surface using a broad knife or scraper
- Repair any cracks by cutting out and filling
- Lightly sand the surface if required with flour paper taking care not to scour the surface and dust down ready for the application of the paint system.

2.5 Preparation of non-porous surfaces

The types of surface to which paint systems may be applied are as follows:

- PVC rainwater goods on external properties
- Perspex, plastic or glass reinforced sheeting
- glass and glazed products
- glazed brickwork or ceramic tiles as found in old school properties
- polished non-ferrous metals.

Consider the following: Most non-porous surfaces do not allow coatings to soak into their structure and gain a grip or lay a foundation for the following system. It is for this reason that the coating has to have the ability to cling to the non-porous surface without movement at a later date which would lead to the breakdown of the applied system. The mediums used in the paints applied to non-porous surfaces are stickier, they give good specific adhesion.

The coatings have the ability to expand and move with the nature of the base material itself. It should be remembered that non-porous surfaces are not generally abraded as the abrasion marks would show through the applied coating looking unsightly.

Non-porous surfaces should be prepared as follows:

- PVC and Perspex goods – clean the surface by degreasing with turpentine, do not abrade the surfaces.
- Glazed tiles – remove all loose grouting and sterilise with a weak solution of household bleach. Degrease the tile surfaces with a solution of sugar soap and warm water then rinse with clean cold water and apply an etching primer. Apply freshly mixed grouting.
- Glass and glazed brickwork – degrease the surface with methylated spirit and etch the glass surface with either French chalk or pumice powder on a damp cloth. This should provide a key for the paint system to be applied without scratching the surface of the glass.
- Polished non-ferrous metals – degrease the surface with turpentine and apply special etching pastes or jellies. These pastes or jellies have the ability to eat into the surface and provide a bond for the coating to be applied.
- The primer used on non-porous surfaces is the final paint coating, for example gloss.

Key point

When preparing non-porous materials remember that any abrasive product that is used can leave scratch marks which will be visible through the applied coating.

2.6 Preparation of surfaces previously papered

Assessment of previously decorated surfaces will be required prior to the redecoration, consider the following:

- Try to identify the surface to which the present wall covering has been applied.
- Has the present wall covering been painted.
- What type of covering is on the surface at the present time, is it anaglypta, blown vinyl, vinyl, ingrain, print etc.
- Is removal of the covering to be carried out using hand or mechanical methods.
- What is the re-decoration to be applied.
- What working practices are to be put in place related to the building's use.

Prior to removing previously applied wall and ceiling coverings, it is considered good working practice to carry out a trial test. This will determine how well the previous paper has been applied and will allow the decorator to determine the method of removal to be adopted. The test will also give up information as to the underlying surface conditions.

The test consists of the following actions:

- If the applied paper is vinyl based, peel off a section of the top layer of paper
- Apply water by sponge to an area of approximately 300 mm square and allow to soak
- After a few minutes use a broad knife or scraper to determine if the paper can be removed
- If this action proves difficult, repeat the process two to three times. This will determine the strategy for the time required for complete removal of paper by hand or determine the need for the use of a steam stripper.

After soaking, if the paper proves difficult to remove, try scoring the surface of the paper with a nail block, scourer and then use a speed stripper. Keep the water as hot as possible and the paper surface continually wet to speed up the soaking process.

(a) Scourer   (b) Scourer   (c) Speed stripper

Figure 2.7 *Speed scourers and strippers*

2.6.1 Tools and equipment to aid removal of coverings

There are many types of hand scrapers available on the market which can be used for the removal of wallpapers. To aid this removal new gadgets have been developed which aid the soaking process. These are devices that scour the face of the paper allowing water penetration and then removal by using the scrapers or speed strippers. Take care when using industrial scrapers or speed strippers that the underlying surface is not damaged. It is easy to remove the plaster or damage the substrate by using too much hand pressure.

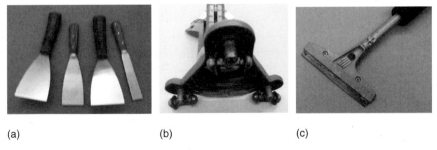

(a)   (b)   (c)

Figure 2.8 *Scrapers and scourers*

If the old wallpaper proves too difficult to remove by hand stripping the use of a steam stripper will be required. This is in effect a large kettle with a device that directs steam to the surface in a concentrated manner.

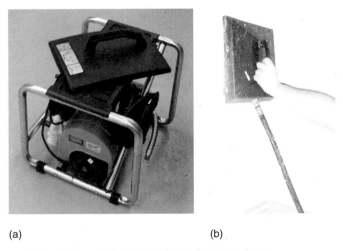

(a)   (b)

Figure 2.9 *Steam stripper and condenser plate*

Before removing any light fittings or loosening switch plates, turn off the electricity at the main distribution box; place a notice on the box to warn others of your actions and to inform them not to reconnect.

Remove electrical fittings and insulate the loose or bare wires. Use the connectors if fitted. It is good practice to indicate which wire is live, which is neutral and which is the earth to enable the re-fixing of items after decoration.

After the protection of loose wires reconnect the supply to the rest of the household.

When decorating around the loose wires switch off the supply, carry out the procedure, remove any adhesive that may be on the wires and switch power back on. Re-fix fittings as soon as possible to reduce the risk of electric shock.

2.6.2 Preparing the designated work area

Prior to actual preparation work, the work area requires preparing for such work to be carried out.

Actions to be carried out to prepare the room for the decorating process are the:

- Removal of furniture, soft furnishings, fixtures and fittings
- The protection of unmovable items such as large furniture and the covering of floors, carpets and decorative features.

The procedures to follow are to:-

- Remove to a safe location all ornaments, soft furnishings, curtains, blinds and pictures
- Dismantle or move system shelving, mirrors, wall lights and remember to replace fixing screws in plug holes to enable relocation of these items after decoration
- Relocate all light items of furniture and move into the centre of the room heavy furniture taking extreme care to cover adequately, remember paint will soak through some dust covers
- It is a good reminder to check if your company is insured for damage that could occur during the decoration processes.

2.6.3 Removal of coverings

Vinyls
Vinyl wallpapers are removed as follows:

- Peel off the top layer by easing an edge and then gently pulling the plastic top layer away from the paper backing
- If the backing paper is in good condition and is firmly adhered to the wall it may be left on the surface as a lining paper. Sand any edges flat at butt joins.

If the backing paper looks to be in poor condition, it should be fully removed. A test that can be carried out to determine condition is to soak an area with water using a sponge. If blisters appear it will indicate the use of a diluted adhesive thus not offering good adhesion to the wall. In this circumstance, any applied paper to this backing could delaminate from the wall after redecoration.

Washable wallpapers
Washable wallpapers are removed as follows:

- Scour the surface with a nail block or scourer to break up the plastic coating. This will allow the penetration of water.
- Apply two to three coats of hot water and allow to soak. You will be able to observe a darkening of the paper as it soaks. If this does not occur the use of a steam stripper will be required.
- Remove the paper with the scraper or speed stripper.

Activity 2.6

Prepare a chart for the removal of wall and ceiling hangings stating the preferred removal method and alternatives. Consider the following headings:

- Type of paper
- Method of removal
- Description of removal

Include the following wallpapers:

- Lining paper
- ingrain
- anaglypta
- super anaglypta
- vinyl
- Blown vinyl
- washable
- lincrusta
- prints
- embossed

Activity 2.7

Write a short report on the importance of the following equipment in relationship to power supply? Log your results in your portfolio of evidence.

| **Figure 2.10** | **Figure 2.11** | **Figure 2.12** |

2.6.4 Procedure to follow for the removal of wall coverings

Most coverings to be removed from walls and ceilings only require the use of a broad knife, scraper, brush, sponge and bucket. The water is applied, it soaks the paper, dissolves the old adhesive and this paper is scraped off the wall. The degree of difficulty depends upon the soaking time.

Key point

When removing wallpapers proves difficult you are not following procedure. Let the water do the work. The most difficult papers to remove are the ones painted after application for this the use of steam strippers will be required.

- Fill a bucket with hot water and add a small quantity of vinegar, the vinegar helps the water to soak into the surface rather than run off it.
- Apply the water to the wall surface in sections and allow to soak. Test to see if the paper is easily removable, if not repeat this process three times before trying to remove the paper.
- Take care not to dig the scraper into the underlying surface when removing the paper and work from top to bottom.
- Once all paper has been removed wash down the walls to remove all traces of adhesive
- Collect and dispose of waste paper in plastic bin liners.

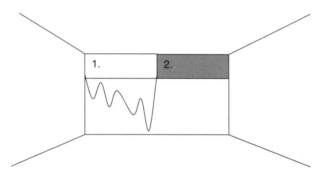

Figure 2.13 *Wall soaking procedure*

Start soaking the wall at the top, the water will run down the wall and penetrate the lower wall thus saving soaking time and making the task of stripping easier.

Steam strippers and their use
- The steam stripper is essentially a large kettle that boils water, they are powered by electricity or gas such as LPG.
- Water is boiled. This water creates steam which is directed along hoses to a condenser plate.
- The condenser plate is held to the wall and the steam softens the paper to be removed. Once the condenser is moved the paper can be removed from the wall using a scraper.
- It is sometimes better to work in pairs if using large industrial steam strippers.

Set up procedure for use
- Check that all parts of the stripper appear to be in good working order and that it has a current portable appliance test date.
- Affix all hoses and condenser plates to the main outlet points on the housing and hoses.

Figure 2.14 *Steam stripper and condenser plate*

- Loosen the filler cap and fill $\frac{3}{4}$ full with hot water. This will heat up quicker and use less electricity.
- Use a funnel for filling and check that the filler cap is firmly replaced.
- Check the side of the unit housing to see if the water gauge is indicating that the boiling container is filled to the correct level. Keep observing this level during use and do not let it fall below $\frac{1}{4}$ full.
- Connect to a power supply and switch on.
- Where possible use 110-volt supply with the correct extension cables and a transformer.
- When the water boils and the steam generates have a bucket available to catch any surges of water through the hoses. This can occur if the unit is overfilled.

Steam stripper safety checklist
When using

- check the water level on the gauge at 10-minute intervals
- when it reaches $\frac{1}{4}$ full disconnect from the power supply, remove the filler cap carefully and refill to $\frac{3}{4}$ full
- connect for reuse. Always be aware that hoses, units and plates become very hot and can scald or burn
- check hoses frequently to ensure hoses do not become kinked
- check connections regularly
- do not trail electric cables away through water
- never point the condenser plate at other personnel
- after use disconnect from the electric supply, allow to cool then dismantle and store correctly.

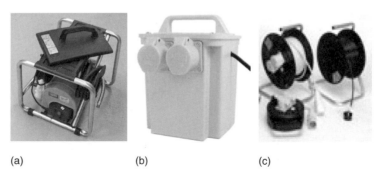

(a)   (b)   (c)

Figure 2.15 *Recommended set up equipment 110 volt*

2.6.5 The next steps to take after removal of wallpaper

After all paper has been removed and the walls have been washed down to remove all traces of previously applied adhesive making good of the surfaces will be required.

Minor repairs
After the stripping process, surface irregularities will become visible such as cracks in the plaster, holes, loose plaster and damaged plaster where the woodwork meets the walls.

- These must be stopped or filled with an appropriate material
- If the repairs are not made good quality decoration is difficult to achieve. Because wallpaper has been removed from a surface, it does not mean that the surface is to be re-papered; it may be specified for a re-paint.

Stopping and filling
The procedure to follow when stopping or filling is as follows:

- Cut out the crack to a firm edge and soak with water
- Apply filler to just below the surface plane
- When dry face fill slightly above the surface plane
- When the filler is dry abrade level.

Test your knowledge 2.5

1. Describe the process of removal of wallpapers by hand.
2. Name the illustrated item of equipment and describe the operational procedures to follow whilst using.

Figure 2.16

3. The above illustrated equipment should be used with a recommended power supply, state the supply.

Place your responses in your portfolio for assessment.

Activity 2.8

Produce a series of diagrams that will illustrate the filling procedure. Attach explanations for each stage. Submit this as portfolio evidence.

2.7 Abrading of surfaces

Surfaces require some form of preparation prior to the application of

- water based paints
- oil based paints
- stains and clear coatings
- specialist decorative treatments
- wall and ceiling coverings.

A combination of preparatory tasks needs to be carried out to enable products to be applied to a required standard, these can include:

- complete or partial removal of previously applied coatings
- degreasing and washing down of surfaces in good condition

- abrading the surfaces with the selected material to achieve the level of smoothness required by hand or powered appliances
- removal of dust and loose particles by industrial cleaning methods.

When selecting an abrasive for the purpose of preparing a surface to receive decoration some knowledge of the following is required:

- what material is the abrasive constructed from
- what is the range available
- how should they be used to best effect and fitness for purpose
- what individual characteristics makes one abrasive a better selection for use than another.

Abrasives can be purchased in the following forms:

- sheet

(a)

- rolls

(b)

- discs

(c)

Figure 2.17 *Types of abrasive*

Or be specially manufactured items such as:

- blocks of fibrous or solid material
- wheels
- stones
- powders
- wire wool
- pads

Abrasive papers – These are classified for use with or without a lubricant to aid the abrading process. If they are used dry, no lubricant is required and they are termed non-waterproof. If they require the use of a lubricant, they are termed waterproof.

Waterproof papers – These are close grained. Non-soluble papers with abrasive particles set on to them with a non-soluble adhesive. The use of a lubricant helps the cutting back process and the paper can be frequently rinsed to prevent clogging of the cutting particle glued to the paper. They are termed silicone carbide papers and are excellent for removing brush marks in previously applied paint.

Non-waterproof papers – Abrasive particles are set in a soluble glue onto a paper backing. They are open grained cannot be used with a lubricant and are generally used for dry sanding to remove bits and nibs from old paintwork prior to re-painting. They do not produce a perfectly smooth brushmark-free surface. The best of its type is aluminium oxide.

Glass paper – Constructed from crushed glass particles which are glued on to a craft paper backing. The size of the crushed glass particles determines its grade. Garnet paper – garnet is a natural mineral and particles of this mineral are glued to a craft paper backing. It is a harder material than glass and is more suited for use on hardwoods. The grain is easily clogged if used on softwoods.

Emery paper – Carborundum is a natural mineral which is glued on to a cloth backing. It is used mainly for the cleaning of metal.

Silicone carbide – A material mix of silica, coke and sand fused together at high temperature. It produces sharp hard crystals and these crystals are glued to a waterproof backing in various grades and used with a lubricant.

Aluminium oxide – Bauxite mineral is glued to a paper backing. A good abrasive which does not easily clog or wear down quickly. It is economical in use.

Key point

If a door surface previously painted was 'ropey' that is to say that brush marks from previous application of the paint coating are visible.

Glass paper would only remove the top profile when preparing for re-painting. Silicone carbide would, when used with a lubricant, remove the profile from the peak through to the trough leaving a perfectly levelled surface.

Figure 2.18
Silicone carbide

Abrasive type	Material	Backing	Glue	Characteristics	Use
Sandpaper	Silica and quartz	Paper/cloth	Water soluble bone glues.	Limited to woodwork use, soft/brittle and not long lasting colour yellow	Dry
Glasspaper	Ground up glass particles	Paper/cloth	Bone glues	Used on dry timber/painted surfaces, various grades available, colour yellow	Dry
Garnet paper	Natural mineral from the garnet jewel	Paper/cloth	Soluble and non-soluble glues	Harder wearing than glass paper, better for use on hard surfaces, colour red	use wet or dry
Flint paper	Ground flint stone	Paper/cloth	Soluble or non-soluble casein glues or resins	A close grained paper used on hardwoods	Wet or dry
Aluminium oxide	Bauxite	Paper/cloth	Soluble bone glue	Disc, belt or sheet form, used by hand, extremely hardwearing, long lasting, economical, colours vary	dry
Silicone carbide Wet and Dry	Fusing of silica and sand together	Paper/cloth	Non-soluble casein glues	Used for fine cutting back of surfaces, close grained papers, black in colour, rinse often	Wet or dry
Emery paper	Carborundum	Cloth	Soluble bone glue	Used on metal surfaces	dry

Figure 2.19 *Abrasive comparison chart*

Activity 2.9

Copy the chart and apply samples of abrasives. Submit as portfolio evidence.

Abrasives are graded as follows:

Silicone carbide – Wet or dry grades   The grading of these papers starts with low numbers, which is a coarse abrasive rising to high numbers which is a fine abrasive. The higher the number the finer the particle size.

Coarse – 80 ➔ 120 ➔ 240 ➔ 360 ➔ 800 – Fine

Coarse (rough)   Fine (smooth)

Initial preparation   Final preparation

Dry abrasive grades The grading for these types of abrasive is identified with either letters or numbers as follows:

Flour – 00 – 1 – 1.5 – 2 – 2.5 – 3 – Fine

Extra fine ⟺ Fine ⟺ Medium ⟺ Coarse

Final preparation   Initial preparation

Paper or cloth-backed abrasives in this category are:

- glass paper
- sandpaper
- garnet
- flint
- aluminium oxide
- emery.

2.7.1 Dry and wet process of abrading by hand

Use only the required amount of abrasive paper for the task in hand otherwise the paper will be wasted. Cut enough to fold into two or three or to fit around rubbing blocks.

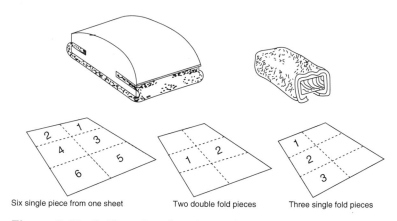

Six single piece from one sheet   Two double fold pieces   Three single fold pieces

Figure 2.20 *Cutting abrasive sheets for use*

Tips to follow when abrading dry surfaces with dry abrasives:

- Follow the direction of the grain of the timber.
- Do not scratch or gouge the surface.
- Abrade in the direction of the longest run of the surface component.
- Use the correct grade of abrasive.
- Change the abrasive when it becomes ineffective.

Key point

110 v/50 Hz is most suitable for use by trainees and students in combination with an 110-v transformer and extension cable.

Figure 2.22

Figure 2.23

- If the abrasive becomes clogged discard it.
- Do not use a lubricant with dry abrasives.
- Wear a dust mask or respirator if adequate ventilation is not available.

Tips to follow when abrading dry surfaces with a lubricant:

- Wet the surface and the abrasive itself, place on the surface and rub in a circular motion in areas of approximately 300 mm.
- Rinse the paper frequently to prevent clogging of the grain.
- Finally abrade in the longest direction of the surface to be painted.
- When abrading ferrous metals use turpentine as the lubricant. Water will cause rusting in the low profile of the metal.
- Use only fine grain abrasive papers on non-ferrous metal to prevent scratching of the surface and rub in one direction only.
- check surfaces before abrading for the presence of lead paints under applied coatings and if there are lead based coatings fully remove the paint to the base material by using paint strippers.

2.7.2 Portable appliances used with abrasives/dry process

Appliances powered by electricity can be used to abrade surfaces quickly and efficiently. The power supply should be either

- 240 v/50 Hz or
- 110 v/50 Hz

(a)   (b)   (c)

Figure 2.21 *orbital/palm/belt/disc sanders*

Key point

If an extension cable has to be used make sure that it is fully unwound otherwise it may heat up and fuse. No electrical appliances should be used by students unless previous training has been given and that training has been recorded in the student's individual training plan.

Electrical equipment set up
- Attach 110-v plug to the power supply cable of the appliance.
- Attach a 13-amp three-pin plug to the supply cable of the transformer.
- Plug in the three-pin plug to a normal domestic supply of 240 v.
- Plug in the appliance two-pin plug to the 110-v transformer socket.

Figure 2.24 *Warning sign*

- If the appliance does not reach the work station, plug a 110-v extension lead into the transformer, fully open and plug in the appliance to the 110-v extension lead.
- The appliance can now be used once the trigger is pressed.

Power appliances with three-pin plugs attached can be used directly with a 240-v power supply but is not recommended as the danger factor is greater.

Follow manufacturers instructions and recommendations at all times when using electrical equipment.

2.7.3 Portable appliances used for wet abrading processes

Mechanical power appliances used for wet abrading are operated by electricity which powers an electric motor attached to a compressor unit which stores compressed air in a receiver. The compressed air is released through air lines which are attached to appliances. The compressed air powers these appliances for the purpose of wet abrading. They are pneumatic power tools. The power supply voltage should be:

- 110 v/50 Hz to the compressor from the transformer
- 240 v/50 Hz domestic supply to the transformer
- compressed air via air hoses from the compressor to the appliance
- this air supply should be regulated to the correct pressure for operational use.

Warning – When wet abrading ensure that the electrical power supply cables to the transformer or compressor are not trailing through water used in the abrading proces.

Types of appliance

(a)   (b)   (c)

Figure 2.25 *Orbital palm sander/rotary sander/sanding pad bases*

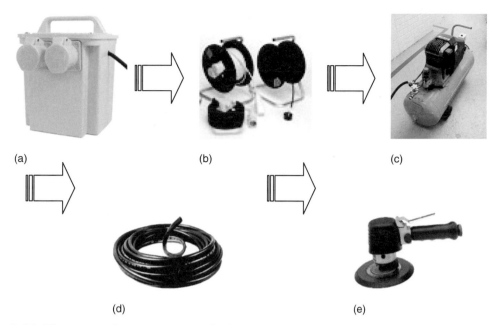

(a)   (b)   (c)

(d)   (e)

Figure 2.26 *The set up from power supply through to the appliance*

Transformer + Extension cable + Compressor + Air hose + Appliance

Pneumatic and electrical appliance information
Rotary disc sander

- pneumatically operated, but requires 80 psi or above for continuous use
- electrically operated, used for dry abrading operations
- various grade of abrasive discs can be obtained for use
- the abrasive discs can be self adhesive or Velcro fitted
- the appliances are mainly used on previously painted surfaces with lubricants
- the sanding head can be changed for fittings that can accommodate moulded surfaces.

Orbital sander

- can be powered by compressed air or electricity
- electrical types consist of a motor housed on top of a rectangular sanding head. The head operates in small circular orbital motions with speeds of 2000–6000 orbits per minute and is

suitable for use on timber, plaster, metal and previously painted surfaces
• a range of abrasive papers, cloths, pads and accessories can be fitted to the sanding heads.

Belt sander

• electrically operated for dry sanding processes
• consists of a motor which drives two cams over which a continuous belt of abrasive paper is fitted
• faster than orbital sanders and is used for flat sanding on timbers and metals
• a range of machines can be purchased with varying belt sizes.

2.7.4 Health and safety at work act (HASWA)

Take the following precautions when using powered appliances.

• wear protective eyewear at all times
• pick up appliances; switch on, use and switch off before laying the appliance down
• maximum operating speeds should be identified on appliances with wheels over 55 mm
• phosphor bronze wire brush attachments must be used in explosive atmospheres
• 110 v/50 Hz transformers must be used on site
• ensure dust extraction procedures are in place
• wear respirators or dust masks
• wear appropriate PPE as required.

2.8
Paint removal using heat

The removal of paint from surfaces is a costly and time-consuming exercise due to the labour required to return the surface to a suitable decorated finish. Prior to the removal of coatings determine which equipment would prove most economical and efficient. The hot air gun or a liquid propane fuelled gas torch.

(a)

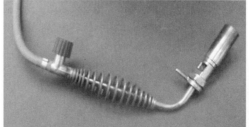

(b)

(c)

Figure 2.27 *Hot air gun/LPG torch/LPG bottle*

If paint has to be removed from surfaces carry out a pre-preparation needs check. Determine if other appropriate preparation methods would be suitable. If the previous coating is badly cracked or is blistering and flaking over most of the surface area then complete removal will be required.

Consider the following before complete removal by heat:

- Will abrading by hand or machine carry out the job of preparation?
- Will dry scraping with a shave hook and hand abrading be enough?
- Will the use of a liquid paint stripper be a more suitable method of removal?
- What is the substrate under the paint system?
- Would the application of heat to the substrate damage the surface itself?

The result of using heat on the substrates listed below is as follows:

Timber surfaces	Scorching
Plaster surfaces	Heat causes cracking and buckling
Brick and stone	Spalling and cracking
Thin gauge metal	Buckling, twisting and holeing
Heavy gauge metal	Heat is conducted away from area

2.8.1 Comparisons of heat removal equipment

Liquefied petroleum gas (LPG)
- For instant lighting, use a spark gun
- Easy to handle equipment where long hoses can be attached for access to upstairs windows on exterior properties
- Controllable flame but can blow out when used externally
- Flame can be concentrated by using various nozzle adaptors
- More expensive than paraffin blowlamps (not often used nowadays)
- Greater risk of fire than with a hot air stripper
- More than one torch can be used from the fuel supply bottle which can be of various bottle sizes
- Possibility of scorching if the flame is set too high.

Hot air strippers
- Electrical power source requiring the use of extension cables or reels
- Should be connected to a transformer using 110 v

Key point

Be aware of the dangers that could be caused by personnel not using the equipment safely. Scorching of the timber can spoil the surface if the appliances are not adjusted correctly, or even fires can be started by setting the timber alight.

- Cannot be used externally with great effect as the outside winds blow away the heat produced by the gun making removal of paint difficult
- Instant heat with variable heat settings
- Light and easy to handle
- Slower than LPG
- Less risk of fire or scorching.

Consider the following before removing any paint coating using heat:

- All flammable material should be moved from the work station, especially when working on windows remove all curtains and nets.
- Remove loose rugs or carpets from around doors to be treated and protect floors with non-flammable sheeting.
- Use a container for disposal of paint debris and have the correct extinguisher in readiness.
- Select the correct hand tools for use with the heat equipment.
- Safe operation of the equipment by trained personnel.
- Check areas frequently that have had the paint removed to ensure they do not burst into flame. Continue for two hours.

2.8.2 General information

- *Wood is a good insulator* – heat will remove oil based paints from timber due to the fact that timber is a good insulator and restricts the heat to the local area only.
- *Metals are good conductors* – removal of paints from metals is not totally effective as the applied heat is conducted away from the local area.
- *Never burn paint off from thin gauge metal* – as twisting, buckling and holes may form in the metal due to heat build up, paint strippers are a better alternative.
- *Plasters, brickwork, stonework* – heat will crumble such surfaces and in many cases cause rippling and delaminating of the finish plaster from the backing plaster.
- *Scorching* – keep the heat source moving constantly to avoid scorching of timbers.
- *Cracking* – use heat shields near glass to prevent cracking.
- *Infected wood* – never try to remove paint films from infected timber as it can easily set alight.
- *Safety inspection time* – finish the process of paint film removal at least 30 minutes prior to and up to two hours before the end of the work shift to facilitate checks for smouldering.

Fire prevention

Types of extinguisher – fire extinguisher cases are coloured red. To identify the contents a colour-coded label is fixed to the case. Black label for carbon dioxide and cream for foam.

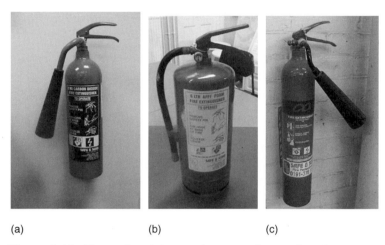

(a)   (b)   (c)

Figure 2.28 *Fire extinguishers – free standing and wall mounted type*

Activity 2.10

Investigate fire extinguishers and produce a chart stating type of extinguisher to be used on various types of fire. Include evidence in portfolio.

Hot air paint strippers

The following are the features of heat gun:

- The common wattage to be found on heat guns is between 1000–2000 W.
- Heat flow and airflow controls are incorporated on the guns with higher wattage.
- 500° Celsius is the temperature setting required to heat up old paint coatings to enable removal of them from the surface.
- Variable airflow settings enable the gun to be adjusted from low to high heat.
- Nozzle attachments are available which direct heat flow spread on to the surface.

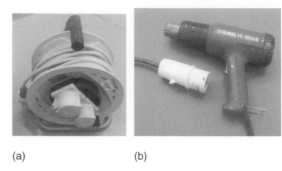

(a)   (b)

Figure 2.29 *110-v extension cable and heat gun*

The following are the types of nozzle available:

- Reducer nozzle directing concentrated heat
- Reflector nozzle
- Flat nozzle spreads the heat in a flat line
- Glass protector nozzle directs the heat away from glass when working adjacent to windows.

A selection of tools and equipment used for the removal of paint coatings by heat are listed below

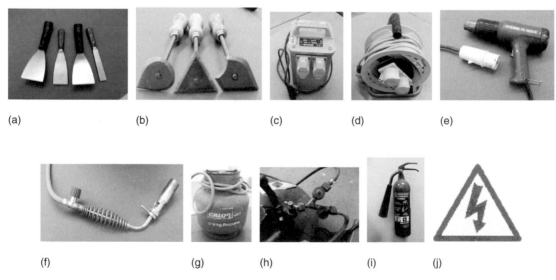

(a)   (b)   (c)   (d)   (e)

(f)   (g)   (h)   (i)   (j)

Figure 2.30 *A selection of tools and equipment used for the removal of paint coatings by heat*

Liquid petroleum gas equipment
Liquid petroleum gas is a fuel source that can be used with gas torches to remove deteriorated paint coatings. Intense heat is given off

when ignited and the flame can be adjusted to control the level of heat directed on to the surface. It is an economical process much cheaper than hot air strippers.

Parts of the LPG equipment consist of:

- A gas supply container. These come in a variety of bottle sizes.

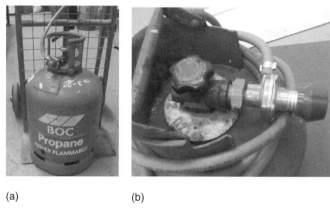

(a)   (b)

Figure 2.31 *LPG supply bottle with transport trolley*

- The regulators which can be adjusted to control the amount of liquid gas that is supplied from the bottle through the hose to the torch.

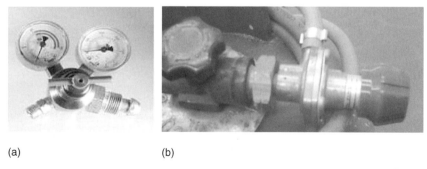

(a)   (b)

Figure 2.32 *Gas regulators*

- Connectors which allow the use of more than one gun to be used from the fuel supply.

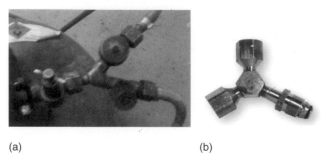

(a)   (b)

Figure 2.33 *Dual connectors*

- Special hoses that are not rotted by the fuel source are used to supply the gas to the torch from the supply bottle. These hoses can be attached by nut onto thread fixings but the snap on and quick release connectors are the best fitting for ease of storing.

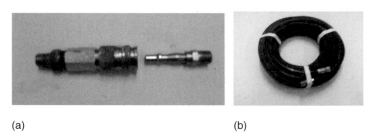

(a)   (b)

Figure 2.34 *Snap connector and hose*

Torches/guns – The torch is a device which directs the supplied heat source to the surface. The ignited fuel at the nozzle can be regulated by a on-off switch on the torch handle.

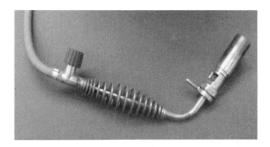

Figure 2.35 *LPG torch*

Figure 2.36 *Double insulation sign*

2.8.3 Comparisons between the operating equipment

Hot air gun
These electrical appliances are double insulated for safety but use 110-v transformers and extension cables wherever possible. This will reduce the risk of electrocution if unforeseen events occur. This is a sticky label, aluminium in colour consisting of one rectangle inside another. Double insulation sign is a safeguard built in to the appliance.

Excellent appliance for the removal of coatings on interior spaces, especially for removing clear coatings. The appliance is not suitable for exterior use as the heat is dissipated by outside temperatures and wind currents.

For safe usage follow these tips:

- Use 110-v supply rather than 240-v direct supply.
- Do not leave connected to supply when not in use.
- All plugs, cables and connectors should be in best condition.
- Any signs of damage to equipment prior to use should be reported and that equipment must not be used.
- Do not direct the heated airflow at any other surface or person when using, it is easy to turn around to speak to someone and in turn burn them accidentally.
- This appliance is not to be used as a drier.
- Never use the appliance in the vicinity of flammable gases or materials.
- Do not lay down the gun on unprotected surfaces when the nozzle is hot.
- Keep body parts behind the nozzle when in use.
- Before storing the appliance let it cool.

Liquid Petroleum Gas Equipment
This equipment is ideally suitable for the removal of coatings both internally and externally. The heat source flame can be adjusted to suit the speed of removal. Take care not to have too high a flame as scorching of the timber could occur. Never use it to remove clear coatings that are to be revarnished.

Lighting procedures to follow for safe usage are as follows:

- Connect the regulator, hose and gun to fuel container if required connect the correct nozzle to the gun.
- Open the valve on the LPG container and adjust the regulator.
- Open the regulator to allow gas into the hose.
- Adjust the gas flow to the gun with the open close control and ignite the gas using a spark gun.

Follow these recommendations when removing paint from surfaces:

- Direct the ignited flame to the surface and allow the coating to soften. As it bubbles move the flame away from the surface and use the hand tool selected to scrape off the coating.

Key point

Caution: Check for leaks on hose and regulator connections. Ensure all connections are tight and hoses are not perished. Keep cylinders away from excessive heat. They can cause tremendous damage if they explode.

- If the flame is held in one place the surface under the coating will scorch.
- Use removal tools that have good edges to ensure clean removal of coatings from the surface.
- If working on vertical surfaces, work from bottom to top as heat rises.
- Follow the applied heat flame with the removal appliance. This will speed up the removal process and keep the surface hot.
- Paint removal will be easier if you follow the grain.
- If you are working next to glass, keep the flame in motion for no longer than the three-second rule and then remove to allow cooling.

The following is the procedure to follow for removal of paint from a panelled door using heat:

- Remove door furniture and store safely for retrieval and replacement upon completion of the task.
- Protect surrounding floor with suitable non-flammable material.
- Have the appropriate fire extinguisher to hand.
- Have a waste container brush and shovel prepared for collection of paint residue on completion of removal task.
- Ensure adequate ventilation during the process.
- Work from the bottom of the door to the top.
- Start with the panel mouldings then the panels.
- Next remove the paint from the bottom, middle, intermediate and top rail.
- Next remove the paint from the muntins, edge and finally the stiles.

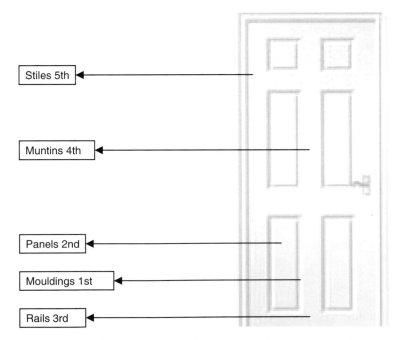

Figure 2.37 *Removal of paint procedure from a panel door*

Test your knowledge 2.7

1. Identify and describe the following components of LPG equipment.

Figure 2.38   **Figure 2.39**

Figure 2.40   **Figure 2.41**

2. How should the risk of electrocution be reduced when setting up hot air equipment for use?
3. When using LPG equipment describe the ignition and extinguishing sequence and the reasons for the sequence.
4. Name and describe two hand tools used for removing the paint from the surface.
5. What safety precaution checks should be made before leaving premises after burning off paint coatings?
6. What sign is fixed to the housing of a hot air gun to indicate it is safe to use?
7. How should liquid propane bottles stored?
8. Name one advantage and one disadvantage of using LPG equipment for paint removal.
9. Name one advantage and one disadvantage of using hot air guns for paint removal.

File your responses as evidence in your portfolio.

2.9 Removal of paint by liquid strippers

Warning – paint removers contain chemicals which are a danger to the operatives health. They can cause burns to the skin and if the fumes are inhaled they will damage the lungs, some can cause cancer. When handling such products read the manufacturers information and follow their recommendations for usage of such a product.

Safety precautions to take when using such products:

- wear goggles to protect the eyes
- wear respirators if fumes are given off
- wear chemical resistant gloves to protect the skin
- wear overalls to protect your clothing
- do not use strippers near any possible source of flame, sparks or heat and never smoke.

Ensure that adequate ventilation is possible. If using paint removers indoors, try to achieve cross ventilation by opening doors and windows.

There are two categories of paint strippers

1. *The solvent-based type* are spirit or water based. The water-based type are adequate for use with most decorative stripping requirements.
2. *The caustic-based type* are alkaline in nature, usually formed from sodium hydroxide. These removers are dangerous if not used correctly. They can darken the wood and usually raise the grain. They are ideal for bulk stripping of timbers if dipped in baths. Avoid skin and eye contact when using caustic alkalis.

Using paint removers removal of coatings is required only when the previous coating system has broken down through the actions of the environment in the form of flaking, peeling, blistering, bleaching or cracking. Remove the coatings by scraping with shave hooks or stripping knives (shown below).

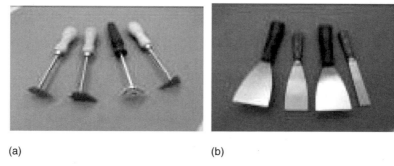

(a)   (b)

Figure 2.42 *Selections of shave hooks and stripping knives*

Paint removers are the ideal medium to use for softening old coatings prior to removal from intricately shaped mouldings such as staircase spindles, dado and picture rails, skirting and architraves.

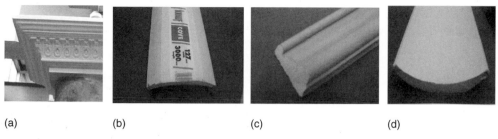

(a)   (b)   (c)   (d)

Figure 2.43 *A variety of mouldings*

Clear coatings such as varnishes and lacquers are removed using strippers from hardwood or softwood surfaces where the surface is to be recoated in a clear finish. The use of heat could result in scorching or timber damage.

Ferrous and non-ferrous metal paint removers are suitable for use on metals for removing old coatings without damaging the surface. Be careful with caustic removers as they could damage some non-ferrous metals by dissolving the metal and pitting the surface.

Glass and glazed products such as window frames, where the use of heat could cause cracking of panes. Strippers do not damage the glass. Do not use paint strippers on surfaces adjacent to Perspex where the stripper would melt the Perspex.

Remember! Ventilation when using strippers.

2.9.1 Comparisons between solvent and caustic paint removers

Solvent remover
Advantages

- enables removal of paint from mouldings
- enables removal of coatings from metals without damage
- enables removal of coatings next to glass.

Disadvantages

- labour consuming
- expensive due to large quantities required if removing coatings from large surface areas
- messy to use thus the need to protect all surrounding areas
- some are flammable
- health hazard.

Caustic remover
Advantages

- economical to use
- non flammable
- effective where components are dipped in vats.

Disadvantages

- raises the grain of timbers and can lift veneer
- never use on aluminium as it will pit the surface of the metal
- causes darkening of the surface
- highly corrosive to the eyes and skin
- will dissolve pure bristle.

2.9.2 Safety precautions

- Wear goggles, gloves and rubber apron to protect eyes and skin.
- Protect surrounding areas from splashing or contamination.

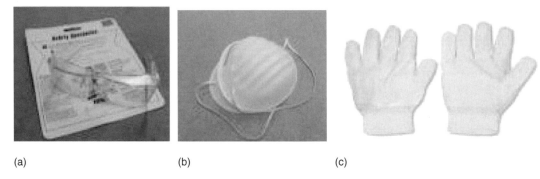

(a)   (b)   (c)

Figure 2.44 *Safety equipment – PPE spectacles/mask and gloves*

Activity 2.11

There are many types of paint strippers available from a wide variety of manufacturers.
A research project to collect information on the

- types available
- usage
- health and safety implications.

would give the decorator valuable information on selection for use to meet environmental needs.

Useful tips on using strippers are as follows:

- Do not smoke, some strippers are flammable
- Release pressure from containers before fully removing the container cap
- Protect the eyes and skin as these removers will burn
- Brush stripper on to the surface liberally
- Allow the stripper to soften the coating. When it wrinkles up it is ready to be removed
- Use the correct hand tools and do not gouge the surface or spoil mouldings
- Never work against the grain
- Use wire wool to remove any residue
- Remove all waste products safely.

Test your knowledge 2.8

1. Why do manufacturers place information on containers?
2. State the items of PPE that should be worn when using paint removers?
3. In what way can paint removers affect your health?
4. State some advantages of using paint removers as against the use of heat methods.

Place your evidence in your portfolio.

Key point

Selection of materials for preparing surfaces prior to the application of decorative products can only come from the following sources:

- Manufacturer's instructions and recommendations
- Experience of working with surfaces and materials
- For a trainee or student, by listening, observing demonstrations and practising preparatory treatments during training or work practice.

If the above are not considered the following consequences could occur:

- The finished decoration could appear unsightly
- The applied decoration will not last the demands required of it by the environment
- The breakdown of the material will not be covered by the manufacturer
- You will lose your customers due to the nature of the produced work and its short life span
- You could lose prospective work by word of mouth.

3 Preparation of materials

3.1 Introduction

3.1.1 Preparation of materials to be applied to surfaces

The preparation of the material to be applied prior to its application is an important consideration, to enable painting and decoration work to be carried out to a standard acceptable to industry and its clientele. Materials that need consideration are:

- paints, water, oil and spirit based paints
- stoppers and fillers such as mastics, putties and ready mixed types
- wallpaper adhesives
- wall and ceiling hangings before use
- abrasive products
- thinners and solvents.

The equipment used to assist in the preparatory processes with the material to be used is an important consideration as well, items such as:

- buckets, roller trays and paint kettles
- plastering hawks, filling or mixing boards
- rubbing blocks
- sieves and strainers (even the use of women's tights is a good cheap alternative)
- paint agitators and stirrers
- container openers
- cleaning cloths.

Appliances used to apply materials should also be checked for suitability, items such as:

- caulking guns, various sizes
- roller poles, varying types.

Key point

If quality painting and decorating is to be carried out, it is important that:

- paints and wallpapers are correctly prepared for use
- the paint to be used is of the correct type
- correct amount has been prepared ready for use
- wallpapers all have the same batch numbers
- the adhesive is of correct consistency
- all products are clean and ready for use.

Finally consider the following:

- selection of the correct paint product
- selection of the correct wall covering for the job
- compatibility of the material with the surface.

Paint should be used from paint kettle's trays or buckets. It is considered poor trade practice to use paint direct from the manufacturer's container. If the manufacturer's container is used

- contamination of the paint can ruin the whole stock
- lids cannot be replaced correctly due to dried paint in the rim
- manufacturers instructions are obliterated by paint coverage.

When decanting paint from the manufacturer's container to your selected container follow these simple pointers:

- Remove any dust from and around the container lid by light brushing.
- Open the lid, using container openers. Below are a selection of specially made openers for plastic and metal containers or you could use an old screwdriver or other similar implement.

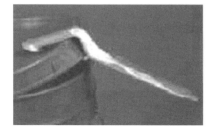

(a)   (b)

Figure 3.1 *Container openers*

Key point

Never use the edge of a paint scraper or filling knife or shears. The blades of knives and scrapers can be damaged by trying to open paint **Figure 3.2** containers.

- Stir the paint thoroughly with a mixing knife until all sediment is dispersed and a smooth consistency is achieved.

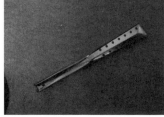

(a)   (b)   (c)

Figure 3.3 *Paint agitators/stirrers*

- Pour the required quantity of paint from the stock into the selected container.

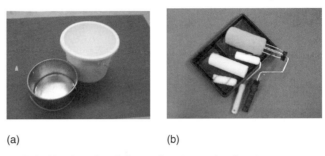

(a)   (b)

Figure 3.4 *Kettle, plastic/metal pots and roller trays*

- Place the stock container on a flat area and clear any paint that has gathered in the rim of the container with a cloth.
- Replace the container lid immediately to ensure the remaining paint does not become contaminated with dust.

Previously used paint containers will require the following actions:

- Clean off dust or dirt from the container lid using a dusting brush.
- Open the container taking care not to damage the rim.
- If paint skins are present on the surface, gently cut the skin away from the edge of the inside of the container.
- Lift out the skin intact if possible and dispose of into a waste receptacle.
- Place a paint strainer on a paint kettle and pour the required amount of paint through the strainer. This will remove any bits of skin or contamination present in the previously used paint.
- Remove the strainer from the kettle and clean or dispose of as necessary.
- Replace the container lid firmly.

Skins form on the surfaces of paints inside containers due to the oxygen or air trapped inside the container when the lid is replaced.

Figure 3.5 *Paint sieve or strainer*

Thinning

New and unused paints that are to be applied by brush do not usually require the addition of solvents as the thinning can lead to the impairment of coating performance. The following are some instances where thinning of the coating is allowed:

- in the preparation of coatings to be sprayed
- for old opened paints that have lost their solvent through evaporation
- to adjust paints to allow penetration to porous surfaces to seal them.

Manufacturer's instructions should be followed before attempting to thin any coatings. Once altered the coating cannot be expected to perform as per manufacturer's specification.

Water based paints such as emulsions can be thinned to aid application but more coats may be required due to the loss of opacity created by the thinning process.

(a)   (b)   (c)

Figure 3.6 *Types of thinner – methylated spirit/turpentine/substitute turpentine*

If paints are to be thinned add only small quantities of correct thinner until the required consistency is reached. For special paints follow the manufacturers instructions regarding thinning.

Key point
The addition of incorrect solvent to a coating will render the coating unusable. The following could occur: • the coating becomes impossible to apply to surfaces with brushes and rollers • the coating congeals • the coating will not dry.

A guide to thinners	correct thinner
Oil based	
Primers/undercoats/gloss and eggshell finishes/varnishes and glazes/ scumbles	white spirit
Water based	
Emulsion paints/vinyl paints/acrylic primers/undercoats and gloss/PVA/ designer distempers	water
Spirit based	
Wood stains/knotting/shellac/ French polish	methylated spirit

Figure 3.7 *Comparison chart*

3.1.2 General information

Paint strainers

Paint strainers can be made from any material that will sieve out any contaminants present in previously used paint or clear coating, such as old skins, dust, grit or dirt picked up by the brush in previous painting processes and then transferred back to the stock container as unused paint.

Strainers can be:

- *Nylon stockings*   An economical way of cleaning up used paint, by straining out the contaminants and then disposing of the used stocking into general waste bins.
- *Disposable strainers*   These cardboard and muslin strainers are cheap to purchase and can be disposed of after use.
- *Metal strainers with changeable gauzes*   Gauze filters can be obtained in various mesh sizes of 40 the coarsest, 60 medium and 100 the finest. When pouring paint into the strainer give the paint time to flow through the gauze. Do not try to speed up the process by agitating the paint and forcing it through the sieve with a knife. The gauze could become damaged or blocked. After use clean the gauze with appropriate cleaner solvent.
- *Vacuum assisted*   This type of strainer requires a supply of compressed air and when set up for use really speed up the straining of paint. It is a lid device which incorporates a venture jet, a funnel and gauze. The lid is placed on top of the container, compressed air is blown through the jet and a partial vacuum is created within the container and paint is sucked through at speed into the holding vessel below.

Paint containers

It is essential that kettles, buckets, troughs and trays that are used to hold paint products during operations, to enable application of coatings, should always be immaculately clean. If old paint is present it could become dislodged from the container, fall into the liquid coating and be transferred onto the surface during the application process causing surface defects. If new containers are not available make sure that previously used containers are clear of any contaminants. Do not use containers that have wet paint present when changing colours or coatings. Check to see that the handles or carrying devices are intact and that there are no holes, splits or cracks in the containers.

(a)   (b)   (c)

Figure 3.8 *Containers*

3.1.3 Personal protective equipment

It must be emphasised strongly that your personal health when damaged cannot be replaced so you should take all possible precautions to protect yourself and others.

Personal protective equipment is available for most tasks and must be used. It is your responsibility to work in a safe manner and not to endanger yourself or others. When preparing paints for use ensure that you take adequate precautions regarding protection of your person. Consider protection for **head, ears, eyes, nose, mouth, body and feet**.

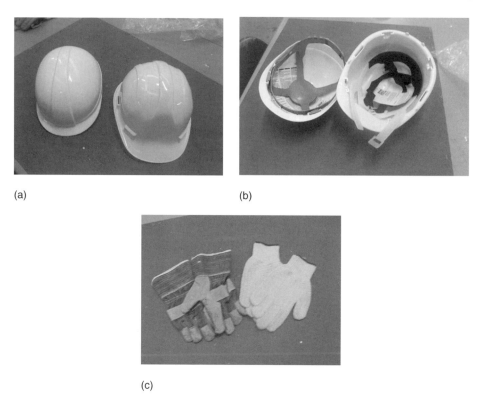

(a)   (b)

(c)

Figure 3.9 *Personal protective equipment – hard hats, bump hats and protective gloves*

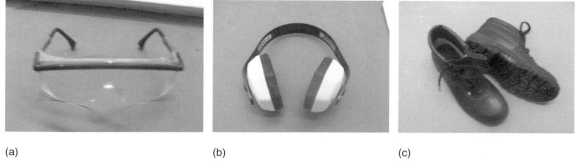

(a)   (b)   (c)

Figure 3.10 *Safety glasses, ear defenders and boots*

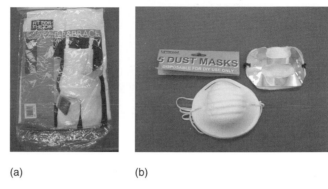

(a)   (b)

Figure 3.11 *Protective overalls – masks and respirators*

Sundries used for decorating tasks require the following materials for preparation:

- powder fillers
- ready mixed fillers
- mastics
- stoppers such as linseed oil putty.

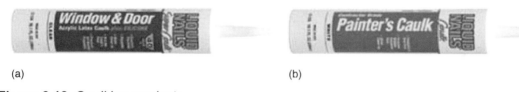

(a)   (b)

Figure 3.12 *Caulking products*

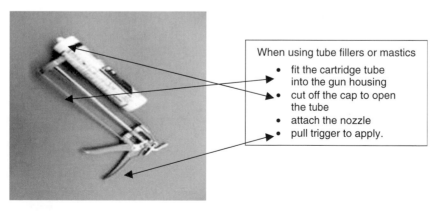

When using tube fillers or mastics

- fit the cartridge tube into the gun housing
- cut off the cap to open the tube
- attach the nozzle
- pull trigger to apply.

Figure 3.13 *Caulking gun*

Key point

- When preparing fillers to use, read the instructions regarding mixing.
- Do not mix too much powder filler.
- Only select the appropriate amount of paste filler.
- Never return unused filler to containers.
- To apply fillers use spatulas/caulkers/filling knives/filling boards.
- After use clean all equipment.

(a) Oil based filler   (b) Putties

(c) Ready mixed fillers   (d) Powder filler

Figure 3.14 *Filler types*

When preparing surfaces using abrasive products a device such as a rubbing block can help produce a flat surface after the abrading process has been completed.

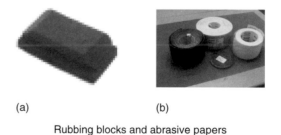

(a)   (b)

Rubbing blocks and abrasive papers

Figure 3.15 *Rubbing blocks and abrasives*

Painting and decorating is a task which uses a wet process to prepare and apply materials. It is for this reason that the operative must take complete responsibility not to damage surrounding property such as fixtures, fittings, furniture and the structure. The operative should be aware that paint splashes, paste contamination all lead to poor quality products. It is due to these work processes that the utmost care must be taken to protect surfaces by sheeting up, masking items and posting warning signs.

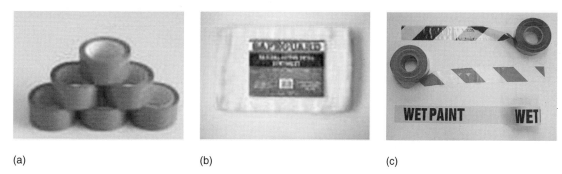

(a)   (b)   (c)

Figure 3.16 *Protective coverings, masking and fixing tapes and warning tapes*

Remember that preparation of materials is just as important as applying the product. Always consider the following:

- prepare the work area
- prepare the equipment
- prepare the material
- apply the decoration
- carry out spot checks periodically during the decorating programme.

The quality of painting and decorating work carried out to a high standard results in customer satisfaction.

3.1.4 Preparation of wall coverings

After purchase, the following checks should take place before opening, cutting and applying the wallpaper:

- are there enough rolls to do the task taking into account wastage
- ensure that the rolls have not been damaged in any way
- is the packaging intact
- are all shade and batch numbers the same
- faults produced during manufacture are not present, check all rolls
- instruction labels are included within the packaging of the rolls or provided as a separate fact information sheet.

LINING PAPER
1000
TRADE QUALITY

Nominal Roll Size: 20.1 m × 56 cm. (11.3 m^2)
22 yds × 22 ins. (121 ft^2)

DOUBLE ROLL

(a)

Lining Paper - Hanging Instructions

IMPORTANT: Before hanging, check that all rolls are undamaged and that you have sufficient rolls to complete the job. If any fault is found, the rolls together with their labels should be returned to the place of purchase. The supplier is NOT liable for faulty hanging. The does NOT affect your statutory rights.

SURFACE PREPARATION:
Ensure all surfaces are clean, firm and dry.
Previously papered surfaces
Remove old paper. Wash down to remove dirt, old adhesive, etc. Fill holes and cracks will filler and allow to dry, then smooth with glasspaper. Size with a thin mix of all purpose strong adhesive.

Emulsion painted surfaces and old plasterwork
Wash surface to remove loose, flaking and blistered paint, using a scraper if necessary, until the total surface is sound.

Gloss/oil painted surfaces and panelling
Wash the surface until completely clean. Fill cracks and seams. Allow to dry, then rub down with coarse glasspaper to provide a key. Lining is essential when you are papering over gloss or oil painted surfaces.

Warning
To minimise hazard in the event of a fire, do NOT hang wallcoverings over expanded Polystyrene veneers and ensure wallcoverings are always firmly stuck down as loose wallcoverings may contribute to the hazard.

HANGING INSTRUCTIONS:
Recommended Adhesive: use adhesive recommended for the final wallcovering. Mix the adhesive, following the manufacturers instructions.
LINING PAPER SHOULD ALWAYS BE HUNG IN THE OPPOSITE DIRECTION TO THE TOP PAPER.
1. Draw a straight HORIZONTAL line on the wall as a guide line to position the first length.
2. Cut the lengths. Allow extra for trimming at both ends.
3. Apply adhesive evenly and ensure that the edges are covered.
4. Fold lengths but do NOT crease and leave each length to soak for the SAME PERIOD OF TIME, until pliable.
5. Hang the first length to the guide line.
   Smooth down with a paper hanging brush or sponge, brushing from the centre to the edges to expel any air bubbles. Trim ends.
6. Hang the other lengths.
   Leave 1 mm gap between lengths. DO NOT overlap.

NOTES:
• For best results, All surfaces should be cross-lined before hanging. This helps to conceal and obscure cracks and blemishes in poor surfaces.
• It provides the ideal preparation for enhancing the final appearance of the new, top wallcovering.
• Alternatively, for people who just want a smooth unblemished surface on which to apply paint, lining paper may be painted with a good quality emulsion paint, when completely dry. This helps to create a more satisfactory decorative effect than painting straight onto a cracked/blemished surface.

(b)

Figure 3.17

Test your knowledge 3.1

1. Name the tools, aids or items that could assist the painter when preparing coatings for use.
2. Explain why it is necessary to prepare paints and materials for use on surface areas.
3. What information is readily available on manufacturer's insturction labels or leaflets? Refer to some available products to access information for this question.

Test your knowledge 3.2

1. Fillers can be purchased in a variety of forms, name three.
2. Name the illustrated item and describe how it helps with the preparation of coatings for use.

Figure 3.18

3. Name the three containers which are used by the painter to dispense paint into.

Figure 3.19   **Figure 3.20**

Figure 3.21

4. Before unwrapping wallpaper for use state the checks that should be made.

4   Preparation of the work area

4.1 Introduction

The painter and decorator who thinks about the activities to be encountered prior to commencement of work will in most cases encounter very few problems. If the job to be undertaken is considered at the pricing or estimating stage, the problems can be planned and more importantly priced for in the cost of the work. There will of course be unexpected snags which are hidden by the existing decoration and a price will have to be negotiated with the customer during the course of the decoration.

If the work is planned carefully actions can be taken that will resolve problems and enable successful completion of a job.

Here are three considerations:-

1. **Liaise with necessary personnel:**
   - the client
   - work colleagues
   - management
   - suppliers.

2. **External influences that can affect the work station:**
   - the client and family
   - suppliers not meeting material delivery
   - the environment
   - the public.

3. **Supporting mechanisms:**
   - specifications
   - timesheets
   - delivery notes.

This chapter will consider aspects that will enable the efficient working practices necessary to achieve quality painting and decorating processes to take place. The end result will be customer satisfaction, recommendations to other persons and hopefully a full working contracts book. In other words a successful company with all its workers in full employment.

In the preparation of the work station the painter and decorator will have to consider actions to be taken

- before the work starts
- during the work processes
- after completion of the immediate work
- before leaving the premises – The actions will include the removal and replacement of components. The responsibility attached to these actions is important as damage can be easily caused to the object or the operator.

Here is a list of considerations which will be explained with further detail later in the chapter. It is not considered good practice to decorate or paint around fixtures or fittings.

Components that need consideration in domestic environments are as follows:

- curtains
- curtain rails
- blinds and track
- radiators
- door furniture
- window furniture
- switch plates and sockets
- carpets
- laminated flooring
- telephones.
- light units both wall and ceiling, mounted and room features such as fireplaces and wall niches.

In commercial establishments other considerations need to be taken. Personal work stations may not be removable to allow decoration to take place, so in these situations it may be necessary to mask and protect items by sheeting them over and working around.

Consider the following when removing or replacing components:

- If you are removing or loosening electrical fittings to work around with tasks switch off the power source at the distribution box and remove the fuse.
- When removing door or window furniture use the correct screwdrivers to ensure the locating screws are not damaged, brass screw heads are easily damaged.
- Store all removed components together in easy storage boxes for later retrieval and refitting.
- Fold curtains and soft furnishings in the correct way. Remember velvet curtains should be pleat folded to prevent creasing.

- Valves should be closed when dropping radiators to prevent water leakage and subsequent damage (if necessary consult a plumber).
- Protect floor areas with appropriate covering material such as cotton dust sheets, polythene strip sheeting, etc.
- Mask immovable items.

4.2 Climatic conditions

Consider the following before attempting to carry out any painting and decorating tasks.

External influences such as:

- rain, snow, sleet or hailstone
- fog, mist or sea fret
- cloudy, overcast light or dark
- pollution
- traffic, both pedestrian and motorised.

Internal influences such as:

- light, dark, dull, gloomy
- poor visibility due to contents of environment
- occupied daily
- internal moving traffic, both motorised and pedestrian
- public areas.

The job location:

- industrial
- coastal
- rural
- internal or external.

4.2.1 The effects of temperature change

Hot, cold and variations of temperature and humidity change can cause the painter and decorator problems prior to, during and after application of products. The results achieved can be the difference between a job successfully carried out and one that has encountered difficulties, resulting in defective or under par work. Weather conditions can spoil work that was effectively executed, but due to the timing of the work was spoiled.

The following problems could be encountered by painters and decorators if the conditions are not favourable:

Too wet (exterior works)

- Inclement weather can prevent contracts meeting completion dates.
- Surfaces are saturated thus preventing applied decoration.
- The film finish can be covered with defects such as flashing, loss of gloss, spotting.
- Adhesion of the applied material does not fully bond with the substrate.

Key point

Wallpaper applied to surfaces in unoccupied properties can remain in the wet state too long due to the walls being saturated with water. Thus the applied adhesive has no option than to over soak the wallpaper causing blistering and delaminating. **Temperature is crucial to enable materials to dry to surfaces, not too cold, not too hot.**

Too cold (internal or external works)

- Drying and curing of coatings will be retarded.
- Application of coatings becomes difficult if not impossible.
- Condensation can mar the finish of applied coatings.
- Wallpapers can over soak, blister and delaminate.

Too warm (internal mainly with the exception of hot summer periods)

- Opacity problems can occur, such as coverage of some gloss.
- Rapid solvent evaporation can lead to application problems.
- Paint fumes can cause problems relating to ventilation.

Too windy (external works)

- Applied paintwork becomes covered with debris such as dust which mars the finish quality.
- The use of access equipment is limited.
- Some preparatory equipment becomes ineffective.

4.3
Customer liaison

To carry out tasks of work it is necessary to co-operate with people to enable the process to have a smooth and successful outcome. There are many roles that can help with this process. Consideration needs to be given

- before commencement of any work
- during the work
- at completion of the work.

The people that you may communicate with could be any of the following:

- client – the customer
- fore person
- charge hands
- other workers from other trades and your own trade
- clerk of works
- site agents
- employers
- office personnel
- suppliers and merchants.

To communicate with such persons will require a level of communication skill. You may have to

- arrange for access to the customers property
- describe what activities will take place during the working day
- programme activities
- ensure good customer/personnel working relationships
- use a chain of command to resolve problems

- co-operate with all interested parties
- seek permissions from third parties (entry to other person's property).

The customer will also require information relating to the following:

- arrival and start times
- break times/lunch times/leaving times on a daily basis
- a forecasted job completion time and date
- progress reports to the customer and employer
- requirements for storage of equipment and materials
- general conservation.

Important considerations related to the working environment and the customer. You may not be aware that what you consider to be normal behaviour is not acceptable to your customer.

- swearing is not acceptable, you may not be aware that you are using such language
- playing of music to loud volumes
- impoliteness and impersonal behaviour
- familiarity with the customer and members of their family
- do not invade others proximity by standing too close, it can threaten!
- personal hygiene and appearance
- requesting refreshments, wait to be asked
- do not help yourself, ask or request the use of comfort facilities
- do not try to hide damage that may occur to the premises, own up and report it to the employer to determine actions required.

Key point

During a job of work the customer may wish to alter the work schedule by changing or adding tasks. If this occurs, contact with the employer should be made to obtain permission prior to this work being carried out. The term used to describe such an activity is a 'variation order'.

Key point

First impressions are important, as this can be the difference between the customer being comfortable with the worker or concerned and asking the employer to remove you from the premises. Remember you are the company's ambassador and future contracts may depend on your relationship during the time spent on the job.

Test your knowledge 4.1

1. What three pieces of paperwork are involved with the requesting, obtaining and paying for materials to carry out decorating tasks?
2. What precautions can be taken to protect the public from hazards and risks whilst work is being undertaken?
3. List problems that may be encountered through adverse weather conditions when organising jobs of work.

Place your answers in your portfolio for assessment.

Activity 4.1

Communication is a necessary activity to undertake to ensure that all persons involved with a job of work know what they are doing. Describe the purpose of the following documents:

- a memo
- a letter
- an order form
- a delivery note
- an invoice
- a timesheet
- a specification
- a schedule.

Use your explanations as evidence towards your qualification. Where possible include examples related to your own workplace.

4.4 Preparation of internal and external work

Working activities are carried out throughout the living environment, be it internal, external, enclosed or in the open, exposed to the elements or protected from them. Whilst the task of work is being undertaken ensure measures are in place to protect the public and their property. Here are some examples:

- If access scaffold has to be erected to gain access to buildings to carry out painting tasks, it will be necessary to protect other users in the vicinity.
- Pedestrian walkways to enable the public to pass below without having to step out on to public highways.
- The hanging of protective covers to prevent paint contamination.
- Safety nets to protect the public from possible falling objects.
- Guarded access ladders preventing unauthorised use.

Strategically placed warning signs are an effective method of warning people of possible dangers in the workplace.

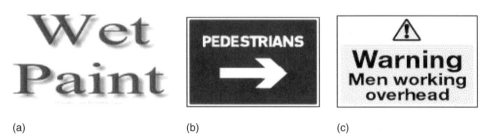

(a)   (b)   (c)

Figure 4.1 *Warning signs*

To create a safe site where the work is carried out without mishap or risk the following actions could be considered:

- Remove debris as necessary, do not let it accumulate.
- Remove any tools, materials, equipment or plant that is not required and may constitute a hazard by being in the way of operations.
- Erect and use appropriate notices.
- Do not use makeshift scaffolds.
- If working at height use safety harnesses and erect safety nets.
- Locate scaffold correctly.
- If possible arrange vacation of premises occupied by other workers.
- Arrange shift patterns to work around normal business use.
- Try not to disrupt other personnel.
- Use or arrange sufficient personnel to efficiently carry out the required tasks.

4.4.1 Preparing external environments for working operations

An important consideration to make is that the jobs of work that we, as painters and decorators, carry out are on other people's properties whether it be private domestic, industrial or commercial.

On domestic properties consider the following when preparing for painting tasks, after all the unaware worker can cause untold damage to the premises.

- In the garden take care to protect lawns, flower beds, shrubs and trees.
- Regarding access protect pathways, gates, driveways and terraces.
- On the structure itself take care not to damage rainwater pipes and gutters, slates or tiles.
- Flat roofs must be protected from damage to the thin membrane.
- Windows and glass especially in conservatories does not look well if covered in minute splashes.
- Storage spaces such as garages and sheds where paints are prepared during the job should be protected from damage.
- Take care not to cause unintentional damage to vehicles.

> **Key point**
>
> **PVC down pipes and guttering**
>
> Remove all guttering from the retaining clips when painting fascia boards; this will enable the application of paint to all of the board thus protecting the timber itself. Many painters, by leaving the guttering *in situ*, miss the top edge of the board and this can result in wet rot eventually.

> **Key point**
>
> **Lawns, flower beds and shrubs**
>
> Whilst using ladders for access to high work, it may be necessary to walk among flowerbeds and on lawns. Take care not to trample down shrubs, flatten bulbs or bedding plants. Ask owner's permission if shrubs have to be removed from walls or to cut them back to enable painting tasks to take place. If you need to place ladders on lawns protect the lawn, and place the ladder on a baton or board. Do not sink the stiles of the ladder into the lawn. The consequences could be dire when the owner sees the damage to his pride and joy.

> **Key point**
>
> **Access**
>
> If you need to access an adjoining property to carry out painting tasks on your job, be sure to ask permission or it could be a trespass situation.

> **Key point**
>
> **Moving of access scaffold**
>
> Take care not to scratch renderings or claddings when moving ladders along a façade. Protect the ends of the tops of the stiles with padding. Do not move ladders by yourself if they are too heavy and cumbersome.

> **Key point**
>
> **Protection**
>
> Driveways, paths, terraces or patios will need protection from paint splashes and preparation debris. If scaffold has to be erected, the floor surfaces could be protected by the use of tarpaulins or even sand, which can be cleaned up after completion of the job. Mask or sheet up brickwork, stonework where spraying of coatings may be required. Keep all gates closed, the loss of family pets or children could be serious. Clean up all work areas after each day's activities and store tools, materials and equipment.

Key point

Consider at all times the following:

- weather conditions
- pedestrians
- traffic, both moving and parked
- the customer and family including pets.

Test your knowledge 4.2

1. When removing small items from locations prior to decoration what precautions should be taken?
2. Prior to removing electrical fittings what should you do to prevent electric shock?
3. When removing curtains and blinds what actions should be considered?
4. How should wooden strip floors be protected during decoration processes?

Place your answers in your portfolio and submit as evidence for assessment when requested.

4.4.2 Preparing internal environments for working operations

When work is to be carried out inside a domestic property many considerations should be taken into account. Before any decoration is carried out a number of pre-tasks need to occur. Consider the following:

- furniture
- soft furnishings
- fixtures and fittings such as lighting curtain and blind tracking
- ornaments
- carpets
- door and window ironmongery
- electrical sockets and wall plates.

If practically possible, remove as many items to other locations to protect and prevent damage during the decorating process.

The following are some suggested procedures to follow to cement good customer relationships:

- On entering premises sheet up all passageways to the point of entry of the room to be decorated.
- Remove all small items such as ornaments to a storage area in a spare room, keep all items together and protected by wrapping in newspaper or cloths.
- Remove all light furniture items to other rooms, heavier furniture should be centrally located in the room and covered with protective covers.
- Curtains and soft furnishings should be removed, folded correctly and laid out in spare rooms to prevent creasing.
- Dismantle curtain rails, fixtures and fittings. Keep all fittings in boxes to prevent loss prior to reaffixing after decoration.
- Remove door and window furniture.
- Switch off electric supply before removing light fittings or to loosen all sockets and wall plates. Insulate all loose wires with insulating tape or connectors.
- Mask up all fixtures such as stone fireplaces, do not use masking tape as the adhesive on the tape can stain the stonework.
- Sheet up carpets if they cannot be removed.

Now you should be prepared to start the decoration process.

4.5 Interpret instructions

Job tasks do not manage themselves. Without the proper planning the smooth running of a contract with the client can be riddled with problems which could lead to company discredit. Plan in advance. The job can then be carried out to specification and managed to the customer's satisfaction. Practical skills are not the only consideration when work is to be engaged, communication skills provide a powerful part of future recommendations. A skilled craftsperson requires the following communication skills:

- The ability to interpret or carry out instructions from various communication devices such as the telephone, the mobile phone, the fax machine, text messaging, emailing.
- Reading of technical information provided by manufacturers in the form of product data.
- Listening to and carrying out instructions given verbally from the following persons, the client or customer, the employer, forepersons, chargehands, clerk of works, technical advisors and others.

Literacy skills may be required from time to time to compile memos or messages to the employer or customer. Calculations for materials to enable simple estimating or pricing of work both on and off site may be required. To be an efficient member of a team flexibility is part of the working life. If you have the ability to meet and resolve problems when they arise, it enables the customer/worker relationship to flourish.

Communication skills will be necessary to be able to deal with the following documentation:

- specifications
- schedules
- drawings
- letters and memos
- order forms, delivery notes and invoices.

You should be aware of such documents and be able to work with them throughout the working contract. A specification informs you of the standard of work expected, how such work is to be carried out and details the work itself in brief. For more detail on specifications discuss with employers or teaching staff.

Painting specification – How should it be written? A specification is intended to provide clear and precise instructions to the contractor on what is to be done and how it is to be done. It should be written in a logical sequence, starting with surface preparation, going through each paint or metal coat to be applied and finally dealing with specific areas, for example welds. It should also be as brief as possible, consistent with providing all the necessary information. The most important items of a specification is as follows:

- The method of surface preparation and the standard required. This can often be specified by reference to an appropriate standard, for example BS 7079: Pt Al Sa21/2 quality.
- The maximum interval between surface preparation and subsequent priming.

- The types of paint or metal coatings to be used, supported by standards.
- The method(s) of application to be used.
- The number of coats to be applied and the interval between coats.
- The wet and dry film thickness for each coat.
- Where each coat is to be applied (le shops or site) and the application conditions that are required, in terms of temperature, humidity, etc.
- Rectification procedures for damage, etc.

A schedule is a list of work to be carried out; it identifies the location, specific area and details requirements.

Paint schedule   This is a chart that indicates the rooms to be painted in a building. The information on the schedule indicates the

- building
- request
- request date
- scheduled to be painted date
- completion date.

Building	Room	Requested	Scheduled	Completion
Rutherford	205	25/06/04	07/07/04	08/07/04
	208	25/06/04	07/07/04	08/07/04
	209	25/06/04	09/07/04	10/07/04
	210	25/06/04	09/07/04	10/07/04
	211	25/06/04	11/07/04	12/07/04
	212	25/06/04	11/07/04	12/07/04
	213	25/06/04	13/07/04	14/07/04
	214	25/06/04	13/07/04	14/07/04
	215	25/06/04	15/07/04	16/07/04

Figure 4.2 *Paint schedule*

Drawings are in the form of plans, elevations and sections. The drawings are to scale and this allows the user of the drawing to take off information to enable the job to be costed for labour and materials.

If you are aware of all the above, your role in the company could be greatly enhanced with the possibility of a managerial role, be it on site or in the office.

Planning of work is an important part of the pre-practical activity and can greatly assist the labour force. Having the confidence to know all the tools, equipment materials and plant are waiting for you gives confidence that time is not going to be wasted with visits

<table>
<tr><td>

Key point

Copies of job sheets should be archived in company storage or retrieval systems to enable future reference to be made if the job is commissioned again in the future. Estimating time, quantities of materials, labour required, times to carry out the work can save time in the preplanning and rescheduling of the work.

</td></tr>
</table>

back to the depot to collect missed items necessary to complete the work.

A job sheet is a useful planner, it can carry information such as

- description of the job
- the address
- scaffold and plant necessary
- materials required
- sundries required
- special considerations
- planned times and dates – start and completion
- signatures of personnel.

Job Sheet

Company address and logo	Unit & element no.		Customer name & address
	Date started		
	Date completed		

Description

Scaffold and plant		Qty	Materials	Qty	Sundries	Qty
Stepladders	small		Emulsion matt/vinyl		Turpentine	
	medium		A.R.P.Primer		Methylated spirits	
	large		Pink primer oil/wood		Cellulose solution	
Trestles	small		Acrylic primer/undercoat		Paraffin	
	medium		Aluminium wood primer		Barrier cream	
	large		Zinc chromate		Abrasive papers/type	
Batons			Calcium plumbate		Putty wood/metal	
Lighweight stagings			Bitumen		Filler powder/oil	
Extension ladders	small		Creosote		Plaster	
	medium		Undercoat white		Adhesive	
	large		Undercoat coloured		Knotting	
Pole ladder			Gloss white		Glue size	
Roof ladder			Gloss coloured		Masking tape	
Mobile tower zip up			Exterior texture paint		Cloths	
Hop ups			Interior texture paint		Sugar soap	
Independant			Eggshell oil paint		Swarfega	
Putlog			Cement paint		Lining papers	
L.P.G. Equipment			Varnish gloss/eggshell		Ingrains	
Hot air guns			Wood preservatives		Anaglypta	
Extension leads			Silicone solutions		Blown vinyl	
Steam stripper			Wood stains		Vinyls	
Special			Special		Special	

Signature(s)						

Figure 4.3 *Painters job planner form*

4.6 Recognising components

This section of the chapter is to reinforce by the use of descriptions and images components that may be encountered when a painting or decorating task is to be carried out.

4.6.1 Shelves and fixing devices

Many rooms require the use of space savers and a common form is the use of brackets and shelving. The normal bracket and shelf can be

Figure 4.4 *Shelves and brackets*

dismantled by unscrewing and removing. Other types have secret fixing points, so investigate before forcing the fitting apart and possibly damaging the unit. The secret fixing types of shelving usually slide away from the wall revealing the fixing point to the wall. Rails such as towel rails, tie rails, racks for shoes all require removal from the space to be decorated to enable ease of access when applying paint or wallpaper. It is also easier to wallpaper and paint if there are no obstacles to cut around.

Place screws back into the retaining holes to ensure you do not lose the location holes whilst decorating. This makes for easy reassembly. Store all fixtures and fitting devices safely to reduce risk of loss.

4.6.2 Grills or ventilation covers

These are constructed from PVC or alloy, and cover openings in walls, ceilings floors or doors. They permit the passage of air for ventilation or heat for warmth as in hot air heating systems. Remove these plates when painting and decorating, and replace immediately. The completed job looks so much better visually. If the grills or plates require painting, make sure that the openings are not compromised, as these grills are designed to let a certain amount of air pass through. If there is gas heating fires fitted in rooms, carbon monoxide poisoning could be caused through blockage of the openings.

(a)   (b)   (c)

Figure 4.5 *Ventilation grills*

4.6.3 Ironmongery

Internal and external doors, and windows are fitted with various fixing devices or furniture to enable their proper use. Each of the devices has its own function which can be for security or visual appearance.

There are main doors to the property, fire doors and shed doors all with their own particular type of fixings. Some are expensive, other fittings can be cheap. Regardless of what is fixed to a door or window, remove as many items as possible apart from the hinges when painting tasks are to be carried out.

- Store all parts safely.
- Take care not to damage screw heads.

- Take extra care with coated brass fittings to prevent scratching to the protective lacquer coating applied during manufacture.
- Clean all fittings before replacement.

Here is a selection of fittings used on doors.

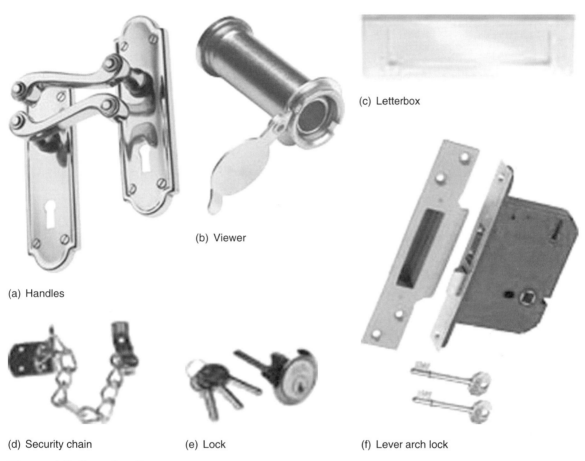

(c) Letterbox

(b) Viewer

(a) Handles

(d) Security chain   (e) Lock   (f) Lever arch lock

Figure 4.6 *Door furniture*

4.6.4 Window fixings

During decoration activities it is usual practice to remove all window furniture, dressings and attachments as they can be easily contaminated with paint products during the preparation and finishing processes. The painter and decorator may require some training on how these items work, the correct removal and replacement sequence. Refer to manufacturers technical data on these products. Do not remove electrically operated curtain mechanisms. It is possible to unhook the drapes and then mask up specialist equipment.

- Remove all soft furnishing such as nets and curtains, and store correctly.
- Release holding brackets and remove blinds, the blinds are removed in the drawn-open position for ease of handling and removal.
- Remove all brackets, holding devices and rails and store safely.
- On completion of decorating tasks replace all fittings as previous and check to see they operate as before.

(a)   (b)   (c)

Figure 4.7 *Detail of window fixing tracks*

4.6.5 Electrical fittings

When loosening or removing ceiling lighting, wall lighting, wall plates, sockets and switches connected to a power supply for the purpose of decorating take the following precautions:

- Switch off the electricity supply at the distribution box.
- Remove the appropriate fuse from the distribution box, be it the lighting circuit or the power circuit.
- Post a notice to warn persons not to replace the fuse and connect the supply as you are working with the power off.
- Tape connectors.
- Store all parts safely whilst working and replace as soon as possible.

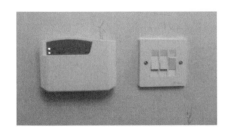  

(a)   (b)   (c)

Figure 4.8 *Selection of electrical components*

4.6.6 Radiators

If radiators have to be loosened or removed it is best to seek the help of a qualified plumber to ensure the householders insurance is not invalidated. Sealed systems can also be damaged by unwitting crafts persons draining off part of the contents of the heating system and not topping up the system when radiators are repositioned.

However, individual radiators can be removed temporarily without the need to drain the full system to enable decoration to be carried out on the wall surface behind. Here is a common procedure:

- Remove the cover caps or thermostats and screw down the valves to isolate the water flow. Note the number of turns required. On re-commissioning use the same number of turns.
- Place enough cloths under the connections to catch spillage when the fixing nuts are loosened. Also have a bowl in place to catch water, which can be black and sludgy if systems have not been maintained regularly.
- Unscrew the cap nut that holds the valve to the radiator adaptor and place the bowl under the opening as you ease the fitting apart. The bowl will catch the drained water.
- Dispose of this water.
- Repeat the process on the other cap nut and lift the radiator off the holding brackets and away from the wall.

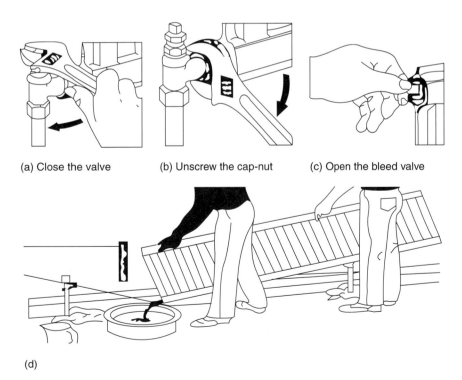

(a) Close the valve   (b) Unscrew the cap-nut   (c) Open the bleed valve

(d)

Figure 4.9 *Radiator removal processes*

- Unscrew the wall brackets, identify the location holes and carry out the decoration.
- Reverse the process to re-commission the radiator.
- Top up the system and bleed to remove trapped air.

4.7 Protection of areas

On commencement of a painting or decorating task, it will be required to prepare the work area by removing contaminants such as general accumulation of dust or manufactured air-borne products from floors, ledges, rail tops, door tops and other unforeseen places where dirt deposits can lay. If this dust and dirt is not removed the applied product can become contaminated giving a less than satisfactory finish.

Use the following aids to help with removal of such contaminants:

- dusters, brushes, brooms, dustpans and shovels
- tack rag and adhesive dusters
- vacuum cleaners.

Brushes, brooms, shovels and dustpans   These can quickly collect and clean up debris such as the removal of coatings, plaster waste after the filling stages to surfaces and wallpaper stripped from ceilings and walls.

Adhesive dusters   These are ideal for removing the fine dust or silt after the abrading process immediately prior to application of paint coatings. This product is muslin coated with a non-drying oil. When used correctly the finished paint film should have no traces of grit or dust present in the applied coating.

Industrial vacuum cleaners   Large quantities of dust and debris created through the preparatory processes of painting and decorating may be collected using these devices.

Trying to collect this debris with a brush and shovel can re distribute some of the waste product.

(a)   (b)   (c)

Figure 4.10 *Dust brush, vacuum brush and pan*

4.7.1 Products for protecting surfaces, furniture, fixtures and fittings

There are a variety of products that can be purchased to offer protection to areas where damage could be caused by the process of decorating. Do not try to use cheap alternatives. The final cost to replace or have professional valeting could be more costly than using good quality protective products at the outset. Look for covers that have hems and eyelets. These covers can be laid down or suspended and are heavy enough to prevent the passage of liquids through their fabric.

Tarpaulin is the name given to waterproof protective covers and they can be purchased with the following specification:

Figure 4.11 *Tarpaulin*

- rubber-coated cotton fabric
- heavy cotton canvas usually green in colour, very expensive
- PVC-coated nylon
- nylon scrim which has been coated with polyester resin.

The common size for purchase is 6 m × 4 m but larger sizes can be made by manufacturers when ordered.

Uses
- Protective sheeting used on and around erected scaffold to protect the public from splashes or falling debris.
- To protect outside work areas from adverse weather conditions.
- To protect equipment and surfaces both internally and externally where heavy pedestrian traffic is encountered.

The above products can, during frosty weather, freeze solid. It is difficult to remove or adjust these covers if frozen.

Dust sheets

The best quality dust sheets should be used wherever possible on domestic properties to prevent the passage of liquids. Heavy cotton twill with hems and eyelets are the best quality. They can be purchased in a variety of sizes, the most common being

- 4 m × 6 m
- 4 m × 4 m
- 4 m × 3 m
- 4 m × 2 m
- special width and length for staircase use.

These covers when used to protect surfaces should be double folded to increase the thickness and offer double the protection to surfaces. They are suitable to cover floors and furniture.

Key point

Cotton twill dustsheets should be taken outside the property regularly during the decorating process, shaken and then replaced. After completion of the decoration process they should be shaken and folded, then stored. When the sheets become impregnated with dirt they should be laundered for reused. Do not forget to sweep up any debris after shaking.

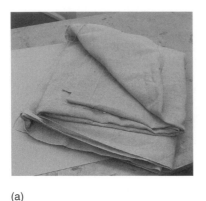

(a)   (b)

Figure 4.12 *Cotton twill dust sheets*

Polythene sheets

This type of covering is used to cover large areas and is used under dustsheets where required. It protects floors where wet preparation processes may be carried out. The sheeting is used to protect floors and is usually cut, fitted and taped in place.

For outside work the sheeting can protect pathways, foliage, outbuildings, etc. from debris created during preparation and application of paint coatings. The polythene can be purchased in rolls of the following sizes:

- 5 m × 4 m
- 4 m × 4 m
- 4 m × 3 m
- 3 m × 3 m
- 3 m × 2 m.

Rolls also come with dispensers.

Masking paper and tapes

Smooth brown paper used for masking and protecting floors furniture, fixtures and fittings can be purchased in rolls of 325 m length with varying widths from 150 to 450 mm. The handling of these rolls is made easier by using dispensing machines.

Self-adhesive masking paper can be purchased where one edge of the paper on one side is coated with a low tack adhesive. It can be purchased in rolls of 50 m in length with widths from 150 to 300 mm. Again use dispensers to apply this type of masking paper.

Crepe masking tape is self-adhesive and can be purchased in rolls that are 325 m long and in widths of 150 mm, 225 mm, 300 mm or 450 mm. This masking paper can be stretched and curved to fit surfaces.

Smooth masking tape, the most commonly used with painters and decorators, comes in rolls that are 55 m long with widths from 12 mm, 19 mm, 25 mm, 38 mm, 50 mm or 75 mm.

Key point

Warning: When using self-adhesive masking tapes, be aware that the adhesive on the tape can stain stonework or furniture. It is good practice to run the fingers down the adhesive side of the tape prior to masking to reduce tackiness and lessen the risk of staining.

4.8 Reinstatement of the work area

Once the painting or decorating job is completed it is not the end of the process. The property needs to be vacated looking better than when you arrived. Do not leave fixtures and fittings for the customer or client to replace. Do not leave debris to be swept up. Create a good impression and recommendations for further work will often follow from existing or new clients.

Follow these simple suggestions:

- Replace all fixture and fittings including curtains: check carpets for damage.
- All tools, materials not used, scaffold and plant should be removed from the site and returned to the depot.
- Make sure there is no paint contamination of floors where you have stored and prepared paint.
- Bag all debris and empty containers, remove from site and dispose of correctly.

To create a good impression and cement customer relationships

- vacuum the carpet
- clean the windows
- clean and polish door and window furniture
- check all items that have been loosened or removed to ensure they are correctly replaced
- in industrial premises ensure that notices, guards, warning signs that required removal for painting processes to take place have been replaced.

Finally, before leaving the premises inform the customer that the work is completed and ask them if they are satisfied with the job.

5 Brushes and rollers

5.1 Introduction

The quality of paint applicators can make the difference between the application of coatings meeting the standards required of industry or the standards accepted by the do it yourself (DIY) sector.

Brushes, like many things in life are accepted for what they are and never a second thought is given to such an item, however they have never been improved upon in many years and they are in reality perfection in design.

Many alternatives have been tried such as paint pads, gloves, spray applicators and rollers, but the brush is still one of the best applicators of coatings. Cheaper alternatives can be purchased but these products are aimed at the DIY market. This chapter explains the qualities of brushes and rollers.

5.2 The construction of brushes

There are four named parts to brush construction:

1. handle
2. ferrule
3. setting
4. filling.

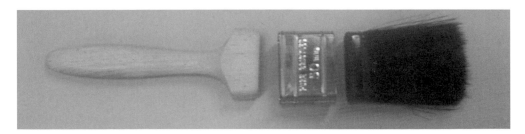

Figure 5.1 *Brush parts*

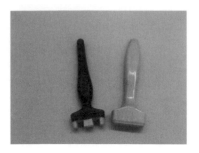

Figure 5.2 *Handle*

The handle   Hardwood is used to construct the handles, blanks of ash, alder or beech are used to shape the handle to the quality of brush required, and these are then sanded and dipped into lacquer to provide a comfortable and hardwearing product.

The filling   Paint brush fillings are:

- pure bristle obtained from the wild boar
- man-made fibres such as nylon
- natural fibres from grassland
- mixtures of fibre, hair and bristle.

(a)   (b)

Figure 5.3 *Fillings*

Most pure bristle is obtained from the Eastern countries such as China, India or Russia. The quality of this bristle, along with the skill of the applicator is what determines the quality of paintwork. The pure bristle should have the following qualities:

- A natural springiness which always retains its shape after each brush stroke.
- A natural taper from root to tip.
- A flag end where the tip of the bristle is split into several strands giving it a soft end which aids application of paint.
- Serrations or fiscules along the entire length of the bristle. This holds paint to the bristle whilst transferring the paint from the container to the surface without dripping.

Horse hair is used in the construction of larger brushes such as wall or emulsion brushes. It is usually mixed with pure bristle and is more hardwearing. Natural fibre is dried grass and plants. Mexican grass (Tampico) obtained from the plains of Mexico. These grasses have no flagged ends, taper or springiness. Natural fibre is mixed with bristle to produce cheap hardwearing brushes for rough surface work. Nylon filaments are used as the filling for new style brushes used to apply acrylics and water based glosses.

Settings This is the term used to describe the gluing of all the bristles together at the root so that bristle loss does not take place during the painting process. There are three materials used for settings:

1. shellac
2. vulcanised rubber
3. synthetic resin.

Ferrules The ferrule holds the handle, filling and setting together forming the brush. The ferrule can be pressed nickel-plated tin for brushes to be used in oil based coatings or copper for water based coatings. Copper is used because it does not rust.

There are two methods of fixing the ferrule to the stock. One is by riveting to the handle, the other is by seaming – terms used to describe the fixing together of all parts of the brush.

> ### Key point
>
> Do not use a brush to apply coatings if the solvent in that coating may dissolve the setting causing hair loss. A shellac setting can be dissolved if introduced to methylated spirits.

(a)

(b)

Figure 5.4 *Ferrule and filling*

The stock

- The bristles are dipped into the setting.
- The setting is heated at temperature to allow curing of the solution.
- The setting is inserted into one end of the ferrule and the handle inserted into the other end.
- The whole stock is then seamed, riveted and pressed.
- The size of the brush and the pure bristle are stamped into the ferrule.
- The filling is then dipped into moth repellent solution.
- Finally the brushes are wrapped for distribution.

> ### Test your knowledge 5.1
>
> 1. What part of the brush has a natural taper, flagged ends and serrations?
> 2. What is the purpose of flagged ends, serrations and fiscules?
> 3. Name the four parts used in the construction of a paint brush.

 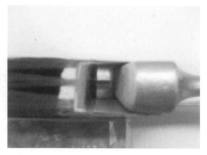

(a)   (b)

Figure 5.5 *Sections through the stock*

5.3
Roller
construction

The paint roller is constructed from the following parts:

- The handle which is constructed from wood or polypropylene is moulded to fit the hand, giving a comfortable grip. Into the handle is inserted the metal arm that attaches the handle to the frame. Good quality roller handles have a threaded internal core to enable a roller arm extension pole to be fitted.
- The frame is a sprung steel cage which is slipped inside the roller sleeve holding the sleeve in place.
- The sleeve and core is the final part of the roller construction. A material is wrapped around a plastic or resin central hollow core. They are spiral bound in some cases for durability. Many types of sleeve suited to the application of most paint coatings are available.

5.4
Types of roller

Rollers are used to apply coatings to surfaces due to the fact that they are cost-effective. Large areas can be covered in a relatively short space of time and at the end of the painting operations the roller sleeve can be discarded rather than wasting time in cleaning the sleeve. Replacement sleeves are inexpensive. Rollers are also extremely useful as they can apply coatings to textured surfaces quickly. There are many types of roller for a variety of tasks:

- single arm rollers
- double arm rollers
- radiator rollers
- pipe rollers
- curved rollers
- decorative rollers
- air fed rollers
- spring loaded rollers.

Roller sleeves A variety of sleeves can be purchased each with a job to do:

- Foam or sponge sleeves can be used to apply gloss coatings.
- Lambswool sleeves in short, medium or long pile are used to apply water based coatings such as emulsions and masonry paints to flat and textured surfaces.
- Synthetic fibre sleeves are a cheaper alternative to the lambswool and are harder wearing.
- Mohair sleeves are used to apply eggshell or gloss paints. They are available in short and medium pile.

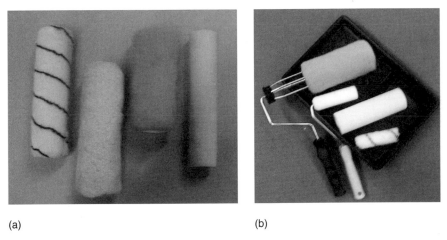

(a)   (b)

Figure 5.6 *Roller sleeves and roller frames*

The roller frame can be single arm construction or double arm construction. The double arm frames hold the larger industrial sized sleeves and are much heavier to handle. The single arm rollers are used mainly for domestic work.

Extension poles can be attached to the roller handles thus reducing the amount of access equipment required to reach ceilings however taking into account cutting in at ceiling and wall angles.

It should be observed that rollers use large quantities of paint and when loaded ready for use can cover vast surface areas. Two operators are usually required, one cutting into angles and one using the roller.

Radiator rollers are used for what they are named after. The roller has a long handle and small sleeve which enables access to the space between the radiator itself and the concealed wall.

The cleaning of rollers after use has to be considered after use. Ask the following questions before making the decision to clean or discard the sleeve.

Figure 5.7 *Details of roller handle and extension pole fitting*

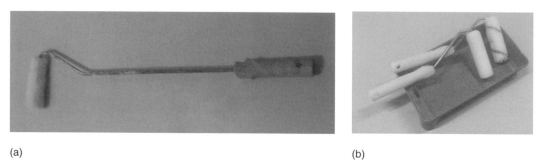

(a)

(b)

Figure 5.8 *Radiator rollers and tray*

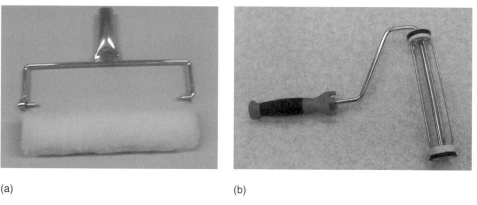

(a)

(b)

Figure 5.9 *Double arm and single arm frames*

- Type of sleeve used – foam and mohair are cheap enough to throw away.
- Lambswool sleeves are expensive and will wash out for reuse.

There are devices which can be purchased where the used sleeve is attached to a spinner impeller which cleans the sleeve quickly. Cleaning sleeves by hand can be time-consuming and costly if using solvent other than water.

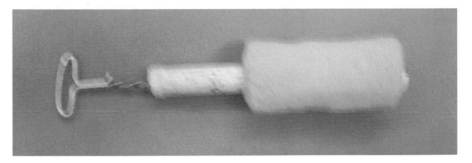

Figure 5.10 *Roller cleaning device*

5.5
Types of brush

There are many types of paint brush that can be purchased from manufacturers, decorators or merchants to carry out the application of the numerous types of paints, stains and varnishes. Some thought should be given to the type of brush selected to enable the coating to be applied to a standard. Do not use good quality bristle brushes to apply new generation acrylic paints, or to apply coatings that may damage the bristle. Special nylon filament brushes can be purchased for the application of acrylics.

Selecting a brush Be careful not to purchase DIY brushes. The brushes do not contain enough filling, have no length out and do not stand up to the rigours of daily use over long periods. Consider the following:

- Is it comfortable to hold?
- Is the ferrule stamped pure bristle?
- Is the filling pure bristle?
- Is there good length out from the ferrule?
- Has it got springiness? The filling returns to its shape when bent back.

Basic set A painter and decorator would include in his basic set of brushes the following brushes. They will allow the application of most coatings to all types of surface. There are other sizes which may be preferred by the individual:

- 175-mm flat wall brush
- 125-flat wall brush
- 100 mm brush
- 75 mm brush   all in pure bristle or mixtures of
- 50 mm brush
- 25 mm brush or sash tool
- radiator brush
- 50-mm oval varnish brush
- selection of acrylic brushes.

Large brushes are used for

- applying adhesive to wallpapers
- applying water based paints to large areas such as ceilings and walls.

Medium-sized brushes would be used to apply coatings to doors and flat surface areas. Small-sized brushes would be used to apply coatings to window frames, mouldings, skirting, architraves, staircase spindles, and others in oil and water based coatings.

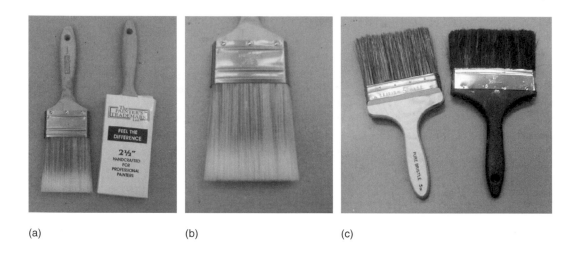

(a)   (b)   (c)

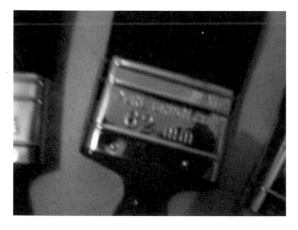

(d)   (e)

Figure 5.11 *Selection of brushes for a basic set*

5.6
Brush and roller maintenance

5.6.1 Storage of brushes used in oil based coatings

- It is an acceptable practice to store daily used brushes suspended in water. Rub out the excess paint on to old newspaper then place the bristles of the brush in water, and water should not touch the ferrule as rusting could occur. The storage of the brush in water prevents the paint from hardening, thus a brush used in white, blue or red, etc. does not require daily cleaning if it is not contaminated with dirt.
- The following day remove the brush from the water, rub out the water from the bristles and rework into the prepared paint.
- The painter and decorator uses paint pots to store like colours together.

Vapour brush keeps can be used to store brushes; it is an airtight container where solvent fumes do the job of the water storage system. Inside the box is placed a solvent bottle and wick. The box when lidded fills with fumes thus preventing air from drying the paint on the brush. When a brush is required, lift off the lid, select the brush for use and replace the lid.

Both of the above storage methods are economical ways of keeping brushes used in paint ready for reuse. One other method used by decorators is to use a mixture of two parts linseed oil to one part white spirit. The brush can be suspended in this solution and it will never go hard. When the brush is used again rinse out in fresh white spirit as the storage solution on the bristles could retard the drying of any applied coating.

Brushes used in varnish should be cleaned out thoroughly after use, rinsing up to three times in the solvent before washing in soapy solution. Then dry and store. If any varnish residue is left near to the setting, the bristles will turn hard rendering the brush useless for use.

Brushes used in water based coatings should be washed out in cold or warm water after use, dried and stored. Do not use hot water as the vinyl in some emulsions could emulsify into strands of soft plastic which is difficult to remove.

Brushes used in specialist paints should be cleaned according to manufacturer's instructions. Do not assume that white spirit will do the job, a reaction might occur which destroys the bristles in the brush, resulting in only one action, to scrap the brush.

5.6.2 General care during use

Brushes and rollers do not require cleaning when short breaks are taken, however they do require protection to retard the drying of the paint loaded on them. Wrap rollers in polythene carry bags overnight or immerse the roller sleeve in the paint being used during lunch breaks. Ensure the roller sleeve is fully immersed or part of the sleeve could dry leaving it unoperable.

Brushes can be suspended in paint or better still suspended in water. Do not let the ferrule become contaminated with paint as the paint will eventually run onto the handle and then on to the operatives hands.

When washing brushes out in solvent take care to spin out all the remaining solvent from the bristles before reworking the bristles into fresh paint. If solvent is left on the bristles, it will combine with the paint and run down the handle, drip and generally become messy to use.

5.6.3 New brushes and care prior to use in coatings

All new brushes are treated with moth-proof repellent during packaging after manufacture. This repellent must be washed from the bristles with warm soapy water prior to use otherwise the repellent will appear as grit and dirt in the applied coating film.

New brushes should also be broken in, that is, to say, wear down the bristles and create a taper on the bristles by using the brush on rough surface work. This will enable gloss paints to be applied and laid off without the risk of runs happening.

Roller sleeves should be selected for the coating to be used. Do not try to apply coatings with sleeves of the wrong type or pile density.

6 Basic coatings (paints, stains and varnishes)

6.1 Descriptions of paints, stains and clear coatings

Paint   A viscous liquid that is applied to built surfaces to enhance the appearance of such surfaces. If constructed surface areas were to be left in their original state, they would not be very appealing unless designed for that purpose alone. Decoration in the form of pigmented coatings is applied for a number of reasons:

- to enable general maintenance
- to preserve
- to make visually acceptable
- to identify
- to give colour.

It would not be very appealing to sit in a leisure or working environment facing dusty plastered walls with foisty smells emanating from the material. Paints offer a fresh colourful appearance to a room which can be colour coded to meet psychological needs for hospitals, restaurants or factories. Paint has to have qualities that enable the applied film to expand and contract with movement of the surface, and last for a given amount of time. Paint has a job to do to prevent costly repair or maintenance.

Stain   These are coatings applied to softwoods and hardwoods that are semi-transparent and allow the figure grain of the wood to show through the coating. There are a variety of types that include opaque or semi-transparent coatings. Some are micro porous and others are purely decorative. Decorative stains are obtainable in colours of the spectrum or as matches to timbers to enhance their natural colour.

Opaque stains are used on cheaper quality timbers to act as protective coatings on fences, sheds and garden furniture – they are termed 'high build' stains. Older type stains such as creosote, coal tar derivative stains are not recommended as they can cause skin damage if the correct PPE is not worn.

Varnishes These coatings are clear unpigmented coatings that offer protection to timber surfaces especially hardwoods where the grain pattern can be enhanced when the coating is applied. There are many types of varnishes such as polyurethane coatings, long or short oil varnishes, yatch varnishes, etc. By their nature they can be very elastic or extremely brittle. The correct varnish must be specified for purpose otherwise it will not offer protection to the surface it is applied to.

6.2 Reasons for painting

6.2.1 *The purpose of painting and decorating*

There are several reasons why paint coatings, wallpapers and fabrics are applied to surfaces, however the main purposes are as follows:

- To decorate and visually stimulate the living environment.
- To protect and enhance the natural life of the surface.
- To identify the surface and give visual information to individuals.
- To offer cleanliness or hygiene to special environments.

Decoration Surfaces such as new plaster, brickwork and timber are building materials used to construct the living environment. Without the use of colour provided by decorative products these surfaces would look dull and drab. There are a vast variety of products available to apply to such surfaces by the painter and decorator on request by the client.

Protection Newly built environments require the application of applied decorative products to prevent such surfaces as timber, plaster, wallboards, etc. from becoming contaminated with dust, dirt and grease. The applied coating allows general cleaning maintenance to be carried out. To some extent it prevents moisture penetration which could cause the surface to deteriorate over a short timescale. The applied coating also allows removal of contaminants such as mould produced by condensation.

Identification The use of shape and colour can be used to great effect in the environment for numerous purposes. Consider safety signage, information giving signs, warning signs, fire fighting equipment in the working and general environments. Contents of vessels and pipelines use a colour-coded system to indicate their contents. Hazards on moving vehicles are identified by black and yellow diagonal lines. Emergency service vehicles are identified by

Key point

Paint carries out an important part of our living environment. Psychological issues can be aided by the sensitive use of colour, especially in children's wards in hospitals where the use of bright colour on surfaces stimulates recovery of the patient.

easily recognised colours. We are conditioned from birth to identify colours with objects.

Cleanliness and hygiene   Non-absorbent paint coatings can be applied to surfaces which enable cleaning to take place on a regular basis. Places like hospitals, laboratories, kitchens which for obvious reasons need to be kept dust, grease and germ free.

6.3
Constituents of paint

Paint composition   Paint consists of a combination of many components, all of which make up the ability of that coating to perform and offer protection to a surface. For the purpose of understanding how paint is produced the following information will simplify the constituents. There are three main ingredients in paint:

1. Pigment
2. Resin or oil
3. Solvent.

Resin   This component is the liquid part of the paint film, sometimes referred to as the medium or the vehicle or binder. The resin gives the paint the ability to stick to a surface and it holds the pigment in suspension in the paint film.

Oil   The older style paints contained drying oils as the liquid part of paint. These paints were more flexible when used externally.

Pigment   This is the colouring of paint and gives body to the paint. It is the solid content. It allows the paint to obliterate the colour of the previous surface to which it is being applied. Pigments are man made or natural. They are expensive to produce and are sometimes mixed with cheaper products named 'extenders'.

Solvent   These are used in paint composition to dissolve the resin (binder) to the correct consistency required of the coating; they do not dissolve the pigment. Solvents evaporate during the drying process of a coating.

Manufactured paints also contain additives to modify their performance on surfaces. Here are some brief descriptions:

Driers/hardeners/catalysts   They are the added agent that allows the drying process to be accelerated changing the paint from a liquid to a solid.

Stabiliser   These are added to a paint to prevent lumps in the paint. They stop pigment from aggregating during manufacture.

Plasticizers   These are added to paints to give a more elastic coating. They prevent the paint from becoming too brittle on drying.

Extenders   These are cheaper forms of pigment added to paints to make them cheaper to manufacture. They bulk out the paint, ease application, allow undercoats to grip the surface and prevent settlement of pigments in the container.

Anti skinning agents   Prevent skins forming in the containers.

6.4
Drying of coatings

Paint, stains and varnishes dry by the following mechanisms:

- evaporation
- oxidation
- polymerization
- coalescence.

The diagrams attempt to show the stages of drying in an annotated form.

Evaporation

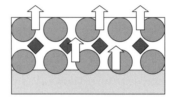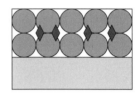

Figure 6.1 *The evaporation process of drying*

The drying process commences once the coating has been applied to the surface. The solvent evaporates into the atmosphere allowing the coating components to settle, dry and form the paint film. If solvent is reintroduced the paint film will revert to its liquid state.

Oxidation

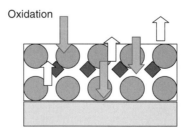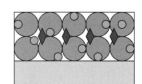

Figure 6.2 *The oxidation process of drying*

The drying process commences with the evaporation of the solvent present in the coating into the atmosphere. Oxygen is then sucked into the paint composition and this oxygen combines with the resin and oils present to form a dried paint film. Reintroducing solvent to the coating will effect no change. The coating has undergone a chemical change.

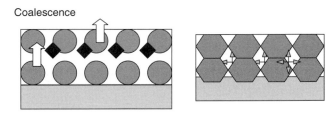

Figure 6.3 *The coalescence process of drying*

The drying process commences with the evaporation of the solvent into the atmosphere. The pigment particles then combine from individual monomers to create polymers eventually forming a honeycomb-like dried paint film.

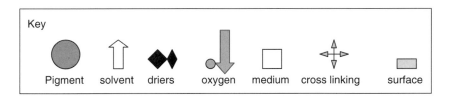

6.4.1 Technical detail of the drying processes

Evaporation
The solvent or thinner known as the diluent escape into the atmosphere thus allowing the remaining components of the composition to dry as a film on the surface. No chemical change has taken place during the drying process. If solvent is introduced to the dried film, it will be reactivated back to liquid form. French polishes, shellac knotting and whitewash all are coatings in this category of paints that dry by evaporation. These coatings are termed 'reversible'.

Oxidation
The process of drying is more complex with oil based paints, when these types of paint films dry a chemical change takes place and the paint cannot revert to its original liquid form. It has converted to another form and is non-reversible. The drying oils and resins present in the composition of the paint attract oxygen molecules into their structure; this creates lynoxyns which internally dry the paint.

Polymerisation/Coalescence
A process of drying of applied paint coatings where oxygen is not absorbed into the film, evaporation of any free solvent takes place then additives in the composition in the form of accelerators and hardeners cause a chemical reaction to occur which results in the film setting. The resin particles in the paint known as monomers combine to become polymers, a complete film. These coatings are

non-reversible and the drying process is sometimes known as curing. A heat reaction usually takes place during the polymerisation drying process. A honeycombing effect takes place with the coalescence process of drying.

6.4.2 The effects of climate on the drying process

Adverse weather effects can create unusual drying patterns to applied coatings, some dry too quickly, others dry too slowly and some do not dry for a few days. This can all be attributed to the following conditions:

- humid atmospheric conditions
- extremely high or low surface temperatures
- high or low air temperatures
- loss of light during the winter months
- the presence of moisture in the atmosphere.

Other influences that can contribute to the failure of an applied coating to dry are:

- the presence of dirt and grease on a surface
- the use of too much thinners or solvent in adjusted coatings
- using non-compatible solvents
- using unprepared paints, not stirring before use.

6.5
Paint systems

Paint systems are combinations of coatings that are applied to surfaces to offer protection and decoration. They are combinations of the following:

- primers
- sealers
- undercoats
- pigmented finishing or top coats
- clear coatings
- stains.

The purpose and function of the paint system is to offer:

- protection to the surface
- the primer or sealer is the foundation of the system and creates the mechanical or specific bond with the surface.
- the undercoat offers the protective build thickness required and obliterates the underlying colour of previous coatings
- the top coats offer a durable flexible film
- clear coatings allow the visible nature of figurative hardwood to be appreciated through a transparent film

- to give colour
- to identify the purpose of the surface, identification of pipelines
- to act as sacrificial coatings
- to prevent vandalism or burglary.

Examples of suggested paint systems

Paint system	First coat	Second coat	Third coat	Fourth coat
Domestic – new unpainted surfaces in oil or water	primer/sealer	undercoat	gloss	
	primer	undercoat	eggshell	
	sealer	emulsion	emulsion	
	emulsion	emulsion		
	special primer	undercoat	top coat	top coat
	sealer	stain	varnish	
Domestic – Previously painted	undercoat	gloss		
	undercoat	gloss	gloss	
	undercoat	eggshell	eggshell	
	emulsion	emulsion		
	acrylic undercoat	acrylic gloss		

Figure 6.4 *Paint system chart*

6.5.1 The function of the layered paint system

The foundation coat of paint on a new surface is the primer or sealer, it forms the key or bond between the surface and the paint. The first coat bonds to the porous surface mechanically (where it sinks into the surface and grips) or specifically (where it lies on the top of the surface) due to the non-porous nature of the surface.

The second coat applied is termed the undercoat. This coat of paint is heavily pigmented and provides high build, and undercoats also have good covering properties due to the high pigment content. The undercoat links to the foundation coat and offers a key to the top coat.

The top or finishing coat offers protection to the system that resists atmospheric or environmental effects for a given period of time. The thickness of a coat of paint varies but on average is 25 microns. The minimum recommended three-coat system should never be less than 75 microns.

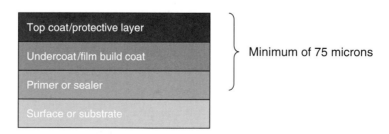

Figure 6.5 *Film build up*

6.5.2 A paint system to use on timber

The graining of previously painted woodwork would require the following system:

- A high build undercoat in the required colour.
- A base coat in the appropriate colour with all brush marks eliminated.
- A scumble glaze (highly pigmented) to match the colour requirement.
- A protective layer of a selected clear coating used as sealer.
- Final top coat of clear coating to offer protection and show off the applied graining effect.

The following is a paint system to use on ferrous metal railings

- zinc phosphate primer
- conventional undercoat to the required colour
- conventional gloss to selected colour or
- micaceous iron oxide undercoat, high build
- micaceous iron oxide top coat.

6.5.3 Selection of paints, stains and varnishes

When selecting paints to use for the paint system use manufacturer's specification for technical information reference. This available information will provide you with

- COSHH technical data
- product data
- HASWA requirements
- spreading capacity in litres
- recoat able periods
- tough dry time
- application methods
- cleaning methods.

6.6
Technical data

6.6.1 Wood primer

Used on softwood and wallboards such as medium density fibreboard, plywood and chipboard.

Composition   It is a high quality general purpose primer formulated to meet the requirements of B.S. 5358. A British standard for low lead priming paints for wood. It is a oil based medium with lead-free pigmentation.

Advantages Provides a strong bond between the surface and the applied undercoat. It has good penetrating properties.

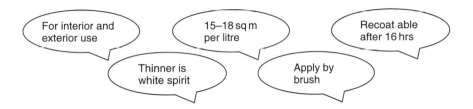

6.6.2 Aluminium wood primer

If timbers have been treated with wood preservatives use this primer. It is also highly recommended for use on hardwoods. The primer has pigment particles that are leaf shaped and when they settle a barrier is formed that effectively seals the surface and prevents bleeding. It is for this reason that it is used as a barrier coating over bituminous or resinous areas.

Composition Top quality aluminium leafing pigment with a specially prepared resin medium is used to produce this primer.

Advantages It offers excellent adhesion to all types of timber and acts as a sealer. It has good flexibility.

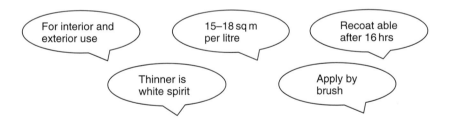

6.6.3 Acrylic primer undercoat

A water thinned lead free wood primer that is especially versatile in that it can act also as the undercoat to build up film thickness. Suitable for internal use but can be used on exteriors. Can be used on plaster but is not recommended for use on metals, plastics or loosely bound surfaces.

Composition Formulated to meet B.S. 5082 and is based on an acrylic emulsion medium.

Advantages Quick drying time with excellent covering power and opacity.

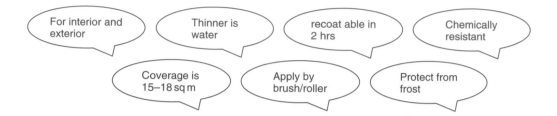

6.6.4 Oil based undercoat

A highly flexible paint which provides coating build to support the top coat in a paint system. It offers good obliteration.

Composition   A non-toxic and lead-free coating, heavily pigmented solid content with specially prepared resin/oil based medium.

Advantages   Excellent covering power providing quick build film thickness to support finishing coating. For internal and external use.

6.6.5 Oil based eggshell

This coating offers excellent performance on ceilings and walls in corridors, canteens, reception areas where resistance to wear and tear is required. Ideal sheen finish.

Composition   An oil based alkyd resin medium incorporating high quality pigmentation. It is formulated to comply with class 'O' flame retardance when tested to B.S. 476 'fire propagation for materials', parts 4, 5, 6.

Advantages   This coating provides a semi-sheen finish which masks surface irregularities. It allows cleaning and resists mild abrasion or dirt contamination. Mainly for internal use.

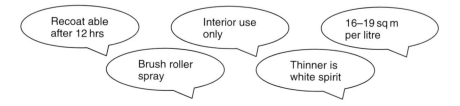

6.6.6 *Alkyd gloss*

A top coating which offers high protection to surfaces if applied according to specification. It has good gloss and colour retention. It is extremely weather resistant and has good resistance to chemicals present in the atmosphere.

Composition Manufactured from the best quality resins and pigments such as oil modified alkyd resins, titanium oxide or lake pigments. Non-toxic with compliance to B.S. 476 class 'O' flame retardance on specified surfaces.

Advantages A tough flexible coating when dry which resists yellowing and chalking. The whites when formulated offer excellent opacity.

Recoat able after 16 hrs | 15–19 sq m per litre | Brush or roller app | Thinner is white spirit

6.6.7 *Emulsions*

There are a variety of emulsion based paint coatings available for use such as:

- vinyl silk emulsion
- vinyl matt emulsion
- contract quality matt and silk emulsion
- designer ranges produced by individual manufacturers.

These coatings are mainly used on interior walls and ceilings but exterior quality types can be obtained for coating external claddings such as rendering or pebbledash. These coatings are termed masonry paints. These coatings should never be used as primers on woodwork or be applied over oil based coatings as adhesion problems will occur. It is not recommended that silk emulsions be applied to uneven surfaces as the sheen of the coating will emphasise the undulations.

Composition These coatings are produced by mixing finely ground pigments in copolymer polyvinyl acetate medium.

Advantages A non-toxic coating which is easy to apply, is quick drying and does not produce heavy paint fumes. Brushes and rollers can also be easily cleaned for reuse.

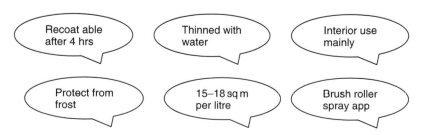

Recoat able after 4 hrs | Thinned with water | Interior use mainly

Protect from frost | 15–18 sq m per litre | Brush roller spray app

6.6.8 Varnishes

There are many types of varnish available for use on surfaces with the aim of protecting those surfaces. The performance specification information offered by the manufacturer should be considered before selecting a varnish for use. Varnishes can be classified as:

Synthetic varnishes include phenolics, alkyds and polyurethanes, all ideal for use on many surfaces; they are tough, clear and quick drying

- Short oil types are ideal for furniture finishing.
- Long oil types are ideal for boats where flexibility of coating is required.
- Phenolics are used in marine varnish manufacture.
- Oleo resinous types yellow quickly through age and look unsightly.
- Polyurethanes are the most commonly used
- Alkyds are the cheapest to purchase.

Copal varnishes include long oil and short oil clear coatings. The short oil varnishes are used internally and the long oil varnishes because of their greater flexibility are used externally. These varnishes are coloured from light amber to clear, but darken with age.

Shellac varnish such as knotting or spirit varnish are made from shellac, dammar and manila resins with balsam or castor oil added as a plasticizer. The mixture is dissolved in methylated spirits to produce a workable consistency. These coatings are ideal for use where

- timber is extremely resinous and the applied coating acts as a barrier
- sealing of surfaces is required to prevent bleeding prior to the application of new paint systems
- a quick drying coating is required.

Water thinned varnishes such as PVA (polyvinyl acetate) varnishes are low odour and practically fume free, they are quick drying and are ideal for use on

- decorative brick and stonework as a protective coating
- wallpapers
- wood panelling and doors
- internal woodwork
- emulsion paints to make them more durable
- as a sealer coat on wallboards.

Two pack varnishes these are unmodified varnishes produced from a combination of polyurethane resin with an added hardener or curing agent. They have excellent resistance to chemical attack, heat and moisture. They have two parts, one the base and the other the activator. Once the two are mixed the varnish drying cycle begins and there

Key point

When using eggshell varnishes it is good practice to apply a gloss varnish prior to the application of the eggshell varnish. If application of the coating is poor, the gloss will show where misses or thin application has occurred. If this is so, an extra coat can be applied.

is only a limited amount of time to apply the coating to a surface. The combination mixture is known in the trade as the 'POT LIFE'.

The main varnish finishes are

- matt
- eggshell
- satin
- satinette
- gloss.

All have varying degrees of reflectance and depend on personal choice.

6.6.9 Stains

Stains can be classified as:

Traditional stains are semi transparent coloured coatings that are applied to hardwoods and softwoods to enhance the grain, these coatings can be obtained as

- spirit based stains
- water based stains
- oil based stains.

The performance requirements of a good stain should:

- be compatible with the surface
- have good light fastness, does not fade
- is quick drying
- have a good spreading rate
- does not raise the grain of the timber to which it is applied.

New age stains such as micro porous stains are replacing traditional stains because they deteriorate quickly through the effect of sunlight and moisture. The new age stains offer the following qualities:

- ultra violet rays from the sun are filtered out so degradation does not take place
- they offer high build
- increased durability and flexibility
- are quick drying
- they contain fungicides to prevent mould attack.

Wood preservatives – these types of coatings are used on external rough woodwork such as fences and garden furniture. There are three types:

1. organic solvents
2. coal tar (wear protective clothing when applying this type of preservative)
3. water borne.

7 Application of coatings

7.1 Introduction

The application of paints stains and varnishes to surfaces can be carried out using a variety of application tools and equipment such as:

- brushes
- rollers
- pads
- gloves
- spray equipment.

This chapter will look at traditional methods of applying coatings by brush, and the techniques used to achieve a high standard of work without the presence of surface application defects.

High quality work is achieved when the correct film thickness is applied by brush without producing runs caused by overloading the surface with paint. An applied full wet coat should be approximately 50 microns thick drying out to 25 microns. The correctly applied film thickness ensures that:

- the coating offers durability
- there is adequate flow on application, not too much and not too little
- this leaves only fine brush marks barely visible
- a longer wet edge time is achieved to enable workability of the coating.

If the applied coating is not correctly manipulated, the following problems could be encountered:

- excessive runs, sags or curtains develop on the surface due to overloading
- poor application resulting in surface defects such as ladders, misses, brush marks, dry edges and flashing

- wrinkling on settling in pools of paint at joins and edges
- poor coverage.

7.1.1 Wet edge time

The wet edge of paint only has a brief time allowance where it remains flexible enough to merge with the next section of applied paint to a surface. If the operative takes too long to apply the sections of paint, the solvent evaporates and the edge of the paint sets off (dries). This then proves difficult to merge the sections of paint and joining lines can be seen. This problem can be encountered when the room temperature or the surface temperature is too high. The operative may have to adjust the viscosity of the paint with a touch of thinners. Solvent evaporates at the same rate, but a full wet coat of paint applied to a surface would have more solvent present than a brushed out coat of paint. This makes application more workable until the laying off procedure has been completed. A term used to describe the unsuccessful joining of sections of applied paint is flashing.

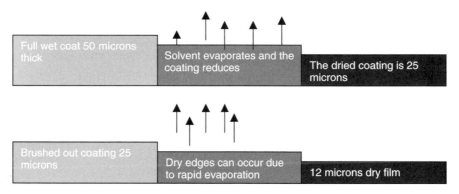

Figure 7.1 *Comparison between full and thinly applied coatings*

7.2 The application of coatings to large surface areas

When large surface areas are to be decorated a number of considerations need to be taken:

- what is the surface area in metres?
- is the surface a ceiling or wall
- is the surface flat or textured
- is the surface porous (absorbent) or non porous
- what type of coating is to be applied, what is the drying time
- consult technical data to decide on method of coating application, brush, roller, spray
- how many operatives will be required to apply the coating.

Key point

Oil based paints require more skill to apply than water based paints. It is better to apply thin coats until confidence can be built up and coatings can be applied that do not form defects due to overloading.

When large surface areas have to be painted some planning will be required to allow continuous application to proceed without the wet edge going off:

- Keep the wet edges as small as possible.
- Reduce ventilation to a minimum whilst applying the paint then increase the ventilation on completion of a full surface. If you can control evaporation by restricting the rate at which it dissipates into the atmosphere you can create a longer working wet edge time.
- Use respirators when reducing ventilation to prevent solvent inhalation.

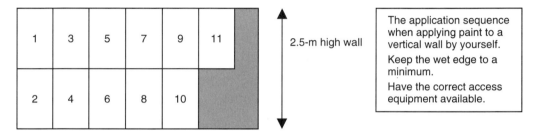

Figure 7.2 *Painting sequence of walls as an individual*

Figure 7.3 *Painting sequence of walls in pairs*

7.2.1 Application of paint to wall surfaces

There are three application processes involved with the spreading of paint coatings to surfaces by brush:

1. The first stage is the transfer from the stock pot using a brush to apply paint to a surface using the criss-cross method (sometimes known as crows nesting).
2. The second stage is to layoff the paint in the shortest direction. This action spreads the paint out into an even consistency on the surface using moderate brush pressure.
3. The third stage is the laying off of the paint in the long direction using the brush. The brush is lightly pulled through the coating to remove brush marks.

At stages 2 and 3 do not load extra paint on to the brush.

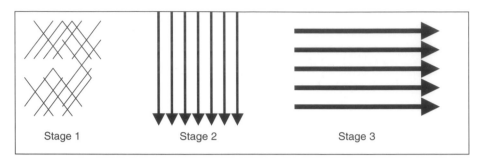

Figure 7.4 *Paint application in stages*

The overlap system   Where paint coatings have been applied in sections there is an area of overlap required to merge the previously applied paint into the presently applied paint. This area of paint will have a wet edge that has started to dry and it is necessary to apply brush pressure at the meeting of the two edges to enable merging and final laying off. If the correct pressure is not applied the meeting joins can be seen as a defect which can result in sags, runs and curtains. The merging of the two wet edges at the stages of application should be between 75 and 150 mm.

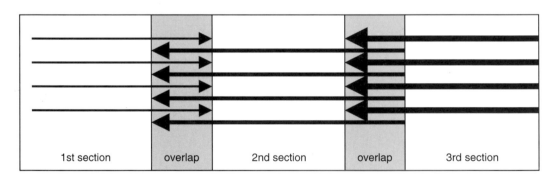

Figure 7.5 *Merging of applied sections of wet paint*

7.2.2 Application of coatings to ceilings

Ceilings are large areas that reflect light. It is for this reason that water based paints such as emulsions are applied to these surfaces. The paint application technique differs to that of applying oil based coatings to walls. If a ceiling bed is not perfectly flat, the last coating that should be applied is one that offers a high sheen which will accent all the imperfections and undulations of that surface. For this reason, matt finishes are applied in the following manner:

- apply the paint liberally with a wall brush
- criss cross the strokes evenly overlapping into the next section
- do not lay off otherwise light will be reflected emphasising undulations on the ceiling bed.

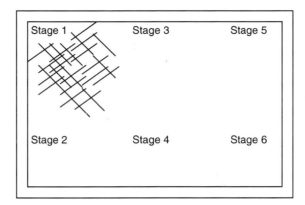

Figure 7.6 *Stages of ceiling painting*

7.2.3 Application of coatings to doors

When applying paint stain or varnishes to doors consider the following:

- Is the door situated in an external or internal location
- Is the door a flush or a panelled door
- Is the door partly glazed or partly panelled
- Is the door constructed from hardwood, softwood, PVC or metal
- Does the door have an individual style, Georgian/Victorian/modern.

All doors are constructed from components such as face sides and edges on flush doors to panel mouldings, muntins, rails, stiles and edges with panelled doors. To effectively apply coatings to these doors there are application procedures to follow which will allow coating standards to be met without application defects developing.

The process of painting tradtional doors – flush doors – is as follows:

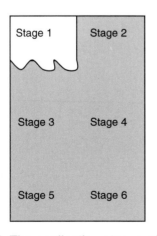

Apply the paint in areas of approx. 300 mm x 300 mm

The edge to paint first on a flush door if you are entering a room and the door opens away from you is the hinge edge.

The edge to paint first as you leave a room and you open the door back to yourself is the latch edge.

Figure 7.7 *The application stages of painting flush doors*

The parts of a panelled door for identification prior to application of coatings by brush:

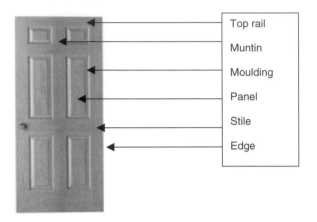

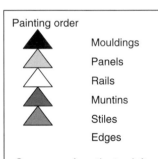

Panelled doors (Interior)

Procedures for painting

Figure 7.8 *Painting procedure of a four-panelled door*

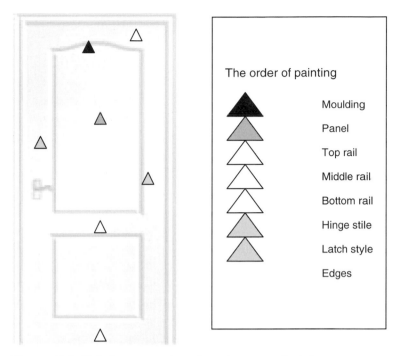

Figure 7.9 *Painting procedure for a two-panelled door*

Painting procedures for more complex glazed door styles

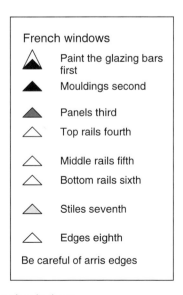

Figure 7.10 *Painting order of French windows*

Figure 7.11 *Painting order of patio doors*

Painting procedures for exterior doors

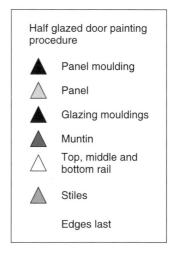

Figure 7.12 *Painting order of a half glazed and panelled door*

Garage doors These require a slightly different approach to painting due to the larger area and the drying time of the coating itself. The painting order of three garage doors are recommended but it is up to the individual to decide on best practice.

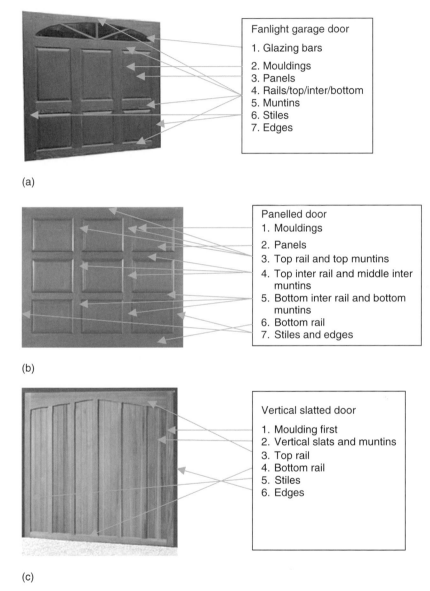

(a)

Fanlight garage door
1. Glazing bars
2. Mouldings
3. Panels
4. Rails/top/inter/bottom
5. Muntins
6. Stiles
7. Edges

(b)

Panelled door
1. Mouldings
2. Panels
3. Top rail and top muntins
4. Top inter rail and middle inter muntins
5. Bottom inter rail and bottom muntins
6. Bottom rail
7. Stiles and edges

(c)

Vertical slatted door
1. Moulding first
2. Vertical slats and muntins
3. Top rail
4. Bottom rail
5. Stiles
6. Edges

Figure 7.13 *Painting orders for garage doors*

7.2.4 Application of coatings to windows

There are numerous types of window design, all with different types of openings. Window painting is probably one of the most time-consuming activities that the painter has to carry out when decorating properties both internally and externally. Consider the following:

- What material is the window constructed from, is it timber, metal or plastic?
- What design features does it contain and how do the opening parts operate?
- Is the window design of the sash type, the casement type?
- Is it a period design such as a Georgian window with narrow glazing bars, or is it a sash window of the Victorian period, etc.?

When all the information has been collated work out a time strategy to enable the window to be painted in the quickest timescale.

Painting procedure recommendations for the following windows:

Painting procedure:

- Open all windows and paint rebates.
- Paint top hinge edge with small fitch.
- Starting from the top paint the glazing putties or beadings, then the flats.
- Work from top to bottom and from side to side.
- Paint the header capital
- Paint the cill.

Figure 7.14 *Bay/bow window*

Painting procedure for sliding sash window:

- Slide the top sash to below the bottom sash to reveal the back horizontal frame of the top sash and paint.
- Paint the sliding rebates.
- Paint the top glazing bars beads and frame.
- Paint the bottom glazing bars beads and frame.
- Paint the bottom sash closer edge.
- Paint the window architrave header horizontal.
- Paint the vertical architrave frame.
- Finally paint the cill.

Figure 7.15 *Sliding sash*

Key point

It should be recognised that all areas decorated by painters and decorators have descriptive names given to enable recognition. Staircases have spindles, baluster rails, bull nose ends and newel posts. Rooms have niches, chimney breasts, internal and external corners. Ceilings have centre pieces.

Research will be necessary or discussion with associates to become fully aware of all components of interiors and exteriors.

7.2.5 Application of coatings to linear work

The application of paints, stains and varnishes to linear work requires the cutting in skill of the painter and decorator to give the finished appearance of a room clean straight line. There is nothing worse than the eye being distracted by undulating colour next to each other.

Linear features include the following room components:

- coving or cornice
- picture, dado or decorative panel rails
- skirting
- door and window architraves, ycasements or frames

When applying successive coats of paint, stain or varnish to such surfaces take care not to overload the surface which could, over time hide the delicate mouldings, ornamentation and enrichments on those decorative surfaces.

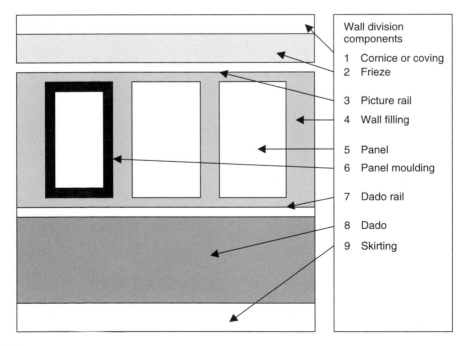

Wall division components

1 Cornice or coving
2 Frieze

3 Picture rail
4 Wall filling

5 Panel
6 Panel moulding

7 Dado rail

8 Dado

9 Skirting

Figure 7.16

Test your knowledge 7.1

1. After application of paint to a flat surface describe the laying off process.
2. List the painting procedure for a panelled door.
3. What defects can develop on a surface if too much paint has been applied?
4. Describe the term 'wet edge' and the problems that can be encountered with it.
5. List the linear components of a room that may require paint application.

File your responses for use as evidence in your portfolio.

8 Application of wall and ceiling hangings

This chapter sets out to explain all the processes involved with the application of wallpapers from preparing the paper and adhesives to the actual hanging. Descriptions of types of papers will also be included. Preparation of surfaces to receive wallpapers was covered in Chapter 2.

The types of papers to be covered are as follows:

- preparatory papers
- semi-finished papers
- finishing papers
- embossed papers and relief papers

The range of papers covered includes:

- lining papers
- ingrain papers known as wood chip
- anaglypta
- supaglypta
- lincrusta
- blown vinyls
- vinyls.

The painter and decorator, on assessment of the surface to be papered, will recommend the use of certain types of paper to mask surface irregularities. These papers are usually embossed or have a relief and a texture or pattern. Pattern books have been produced by manufacturers and offered to retailers to enable the decorator to show customers the available range; the customer can then select their choice.

The selection process needs to be pre-planned. The pattern book is booked out from the retailer and delivered to the customer who in turn makes the selections. The decorator checks with the retailer that the wallpaper selections are available or if ordering is required from the manufacturer. Dates can then be arranged for delivery and for the decoration to take place. In most cases selections can be made off the shelf for the more standard designs and patterns.

Figure 8.1 *Wallpaper pattern books*

8.2
Types of coverings

Preparation papers   These are papers that are applied to bare surfaces to offer a clean working surface on which to apply the finishing wallpaper, or to provide a textured surface to which paint coatings can be applied.

The purpose of the preparatory paper is to offer the following:

* a surface that has even porosity
* a surface that hides undulations and irregularities
* offers temporary protection from dampness
* enhances the visual effect of open weave Hessians (advanced wall hanging)
* offers the undercoat paper to receive the top paper
* offers insulation.

Lining papers   These are plain pulp papers with one side slightly polished. They are available in various thicknesses or weight known as grades. The grades can be obtained from 800 lightweight to 1400 heavyweight. The heavier grade lining papers are for poor surfaces or where colour is to be applied as a finish coat.

The roll width dimension of lining paper is slightly wider than wallpaper at 550 mm as against 530 mm.

* 800 grade used for standard application to walls and ceilings
* 1000 grade is a medium weight lining paper used on old and new plaster walls and ceilings with minor imperfections (weight: 150 g per square metre)

Key point

Remember – The preparation paper has to be hung correctly to high standards to offer a good base for the finishing paper. It is vital that the preparation of the surface is of a high standard otherwise poor preparation will show through both the lining paper and the top paper. Compatible adhesives are required to ensure no reaction takes place after the papers are hung.

- 1200 grade is a heavyweight lining paper used on old and new plaster walls and ceilings with moderately rough surfaces (weight: 170 g per square metre).
- 1400 grade is a super heavyweight lining paper used on old and new plaster walls and ceilings with rough surfaces, giving exceptional strength (weight: 190 g per square metre).

8.2.1 Types of lining papers available for use on surfaces

Brown lining
This similar to wrapping paper which is used under Kraft backed coverings such as lincrusta, 90 gm/m^2 standard roll.

Pitch paper
This is a brown lining paper which is coated on one side with pitch. These papers are used as a temporary barrier on walls that suffer from penetrating dampness. The joins must be overlapped and feathered when hanging to prevent moisture seeping through. The paper should also project 300 mm in all directions past the damp patch.

Cotton or linen backed reinforced lining paper
This is a heavy grade quality white lining paper backed with a cotton bandage scrim which is used on surfaces that are badly cracked or over tongue and groove boarding. The scrim side is pasted and hung to the surface.

Metal foils
These are thin sheets or rolls of non-ferrous metals which are applied to walls to act as a barrier against dampness. There are various types available:

- traditional lead foils, considered the most reliable
- aluminium foil, for smaller areas. These foils are very fragile and can be attacked by acids and alkalis present in surfaces
- aluminium coated with thermoplastic adhesive, this type can be ironed on to the wall.

Expanded Polystyrene
A manufactured foam resin which is cut into rolls, sheets or tiles or moulded into coving, cornices and niches. The material increases insulation in rooms where it is applied and can to some extent combat condensation. When applying this material to surfaces be aware of static electricity. Earth yourself prior to touching other people or you may give them a slight electric shock.

All of the lining paper products mentioned are not used extensively but offer a solution to problems that could be costly to repair if a builder has to be called in. Some of the materials may be difficult to obtain from suppliers.

(a)   (b)   (c)

Figure 8.2 *Lining paper. 1400 grade/fibre reinforced/expanded polystyrene*

Key point

All relief decorative wall coverings carry a raised pattern on the face side and are excellent for masking surface imperfections. They offer a wide range of textures and patterns for both ceilings and walls.

Decorative papers   There are wallpapers that come under the category of relief decoration. These papers have a raised pattern or emboss. Some require the application of paint to complete their decorative effect.

Semi-finished reliefs include:

* ingrain
* anaglypta and supaglypta
* vinyl and blown vinyl
* lincrusta.

They can be classified as either solid or hollow relief. The solid relief papers have a flat backing to which the adhesive is applied. The hollow reliefs are embossed; this makes the application of adhesive more difficult.

* The solid reliefs include the blown vinyls and lincrusta coverings.
* The hollow reliefs include the anaglypta and supaglypta coverings.

Ingrain or woodchip
A semi-relief wallpaper manufactured by placing small wood chips or vermiculite between two layers of pulp which are then duplexed together. These papers are produced by countries such as Sweden, Germany the USA and the UK. Each has its own grading from light to heavy particle surface texture in appearance.

* roll size, 10.05 m long × 0.550 wide
* single, double, triple and quad rolls can be purchased.

Anaglypta
Two layers of chemical wood pulp are duplexed together and embossed whilst in the wet stage. This type of paper requires a soaking time after the adhesive has been applied prior to its application to a surface. Care must be taken not to over soak the paper, or over brush when applying otherwise the emboss could be flattened. It is an ideal product to use if surface irregularities require masking.

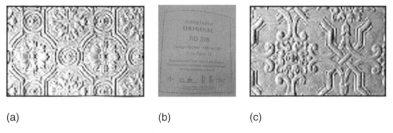

(a)   (b)   (c)

Figure 8.3 *Anaglypta sample patterns/design number stamp on reverse*

Supaglypta

These are produced by embossing a paper prepared from cotton fibre, rosin size, china clay and alum whilst in its wet state. Supaglypta is much heavier wallpaper and has a higher relief in texture than anaglypta. After the application of adhesive soak each length for approximately 15 minutes before hanging to a surface. Each piece must be soaked for the same length of time otherwise matching problems will be encountered due to expansion of the paper. Do not over smooth when applying the paper to surfaces as the pattern could be flattened.

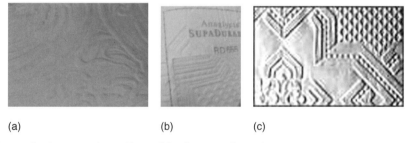

(a)   (b)   (c)

Figure 8.4 *Supaglypta sample patterns/design number stamp on reverse*

Relief vinyls

Commonly known as blown vinyls. These are paper backed polyvinyl chloride products and have offered an alternative to anaglypta and supaglypta which require much more skill to apply. They offer good thermal and sound insulation and have a permanent emboss which cannot be flattened on application. Some are flame proof. These papers are manufactured as semi finished or coloured.

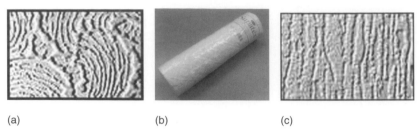

(a) (b) (c)

Figure 8.5 *Blown vinyl sample patterns*

Key point

Comparison – When removing blown vinyl just peel away the vinyl from the backing in the dry state, then remove the remaining lining by the normal stripping method.

Anaglypta and supaglypta will prove much more difficult to remove and might require the use of a steam stripper. Labour is saved when blown vinyls are removed.

Lincrusta

A highly embossed wall covering which simulates wood, stone, rope and brick effects. The face covering material is manufactured by using a linseed oil/putty composition layered on to a kraft paper backing. The texture or pattern is pressed into this composition whilst in its wet state. It is a heavy hardwearing and durable wall covering.

These coverings can be purchased ready finished or in their natural state which then allows for the application of decoration either as a paint system or as a decorative paint system such as glazing and wiping. It takes great skill to apply these coverings to best effect, not recommended for the DIY market. They are very expensive to purchase.

Lincrusta can be purchased in rolls, panels or as borders specially matched to period designs. A protective strip called a selvedge is trimmed off the roll prior to hanging to surfaces. Its purpose is to protect the roll edges from damage during distribution.

Vinyls

Vinyls are manufactured by bonding a hot vinyl coating on to a cotton backing paper. The colours and designs are fused into the vinyl during the process. These products are washable and are not damaged when

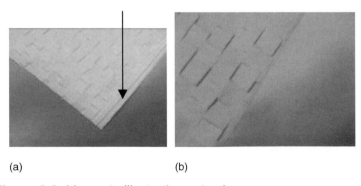

(a) (b)

Figure 8.6 *Lincrusta illustrating selvedge*

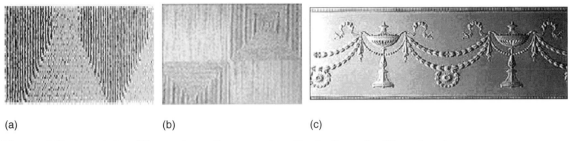

(a)   (b)   (c)

Figure 8.7 *A variety of lincrusta patterns and an illustration of a frieze border*

Activity 8.1

Collect samples of wall coverings mentioned in this chapter and produce a chart that offers the following information:

- product name, designer and manufacturer
- hanging instructions
- adhesives recommended for use with this product
- dimensions, roll sizes available
- cost.

Use this information as evidence for assessment.

Test your knowledge 8.1

1. What is the function of a pattern book?
2. Name the preparatory papers and describe their use.
3. What do the initials R D describe?
4. What preparatory paper should be selected for use on surfaces that are badly cracked?
5. Identify two papers that can be removed from surfaces in the dry state.
6. What does the term 'selvedge' refer to?
7. What grades of lining paper are available for use on surfaces?

Place this evidence in your portfolio for assessment.

cleaned with mild detergents. Strong sunlight will fade the ink colours of some vinyls. When redecoration is required the top layer is peeled away from its backing leaving a lining base.

Technical information: The colours used in the manufacture of vinyl papers are brighter and more intense. The spirit based inks give the colours their intensity. To obtain the fusing effect of vinyl and coloured inks, the printed vinyl is heated to almost melting point by passing it under an infra-red heating element.

Some vinyls have an emboss effect which is achieved by passing the vinyl whilst still hot through embossing rollers. The process is called 'hot embossing'. Vinyls are manufactured using the 'intaglio' method of printing on a rotogravure machine.

8.2.2 Basic adhesives information

Decorative coverings adhesives are, generally known as wallpaper pastes and are available as:

- ready mixed tub pastes in light, medium or heavy grades
- in sachets or boxes, in powder crystal form just waiting for the water to be added
- in tubes or small plastic containers of varying sizes ready for use.

Traditional adhesives consisted of:

- hot and cold water flour pastes
- cold water cellulose pastes
- ready mixed pastes for heavier papers and fabrics
- impact adhesives for vinyl on vinyl or for repair work.

Some understanding of the nature of the adhesive must be observed before a particular type is selected for use with a wall or ceiling covering. Prior to selecting or using any adhesive read the manufacturer's instructions on various packets or sachets.

This gives information such as:

- the types of wallpaper that the adhesive can be used with
- mixing and special instructions
- spreading rates.

Consider the following:

- Is the selected adhesive compatible and will it adhere the paper to the surface?
- Is there the possibility that the paste could stain and damage the paper?
- Are fungicides present in the adhesive to prevent mould growth?

Cellulose adhesives

In paperhanging, cellulose has the highest water content of any paste in general use (around 97%). It is supplied in boxes and sachets and is packaged as a white powder. It is mixed with cold water on the job and can be used with a variety of lightweight materials such as preparatory papers, light relief papers and all vinyls. Its adhesion is mostly of the mechanical type. It leaves very little solids behind and is not suitable for many wall coverings which require greater amounts of initial tack and holding power.

Starch adhesives

Also known as flour paste. Common wheat flour is the most frequently recommended flour for making wallpaper paste. It is supplied in sachets. The user simply adds the powder, sifting it slowly into cold water, while stirring with a stick or whisk to avoid lumps. It is commonly known as cold water starch paste.

Wheat has the next highest water content of any paste in general use, ranging from 90 to 95% water, depending on how much water is mixed with the product. The introduction of vinyl and vinyl-coated paper to the wall covering industry has led to a decline in the use of wheat paste.

Wall coverings that can be successfully hung with starch flour paste include lining paper and residential papers. Most modern adhesives, being either starch or cellulose have fungicides incorporated to prevent the growth of mould.

Border adhesives
Ideal for hanging pre-pasted borders and wallpaper to painted walls and vinyl wallpaper, they are ready mixed, easy to apply and save time. The open time allows more time for positioning and pattern matching.

Specialist adhesives
PVA adhesive
A ready mixed thick liquid paste or glue that offers strong adhesive strength for the fixing of vinyl fabrics, polystyrene sheets, rolls and tiles.

Lincrusta glue (rubber glue)
This adhesive has very strong bonding properties and is a specially formulated adhesive for the application of Lincrusta. The contents of a 1-litre tin are sufficient for approximately one 16.5 m length of Lincrusta.

Stir the adhesive before use and never thin. Lincrusta material should be sponged generously on the backing with warm water and allowed to soak for 20–30 minutes before application. This allows the material to expand and become pliable thus assisting the application process.

Prior to applying the adhesive wipe the back of each length thoroughly with a dry cloth to remove any surplus water. Apply the Lincrusta glue with a general purpose 3-in. paint brush, place the length into position and smooth down with a cloth pad or 4-in. rubber roller using maximum pressure.

Impact adhesives
These are adhesives based on rubber/neoprene. Hazardous vapours are given off by impact adhesives. Refer to COSHH when using this product. Points to remember when mixing or using adhesives:

- Make sure all mixing utensils are clean.
- Follow manufacturer's mixing instructions.
- If you over thin an adhesive, discard it.
- Ensure there are no lumps in the adhesive when ready for use.
- If you are using adhesive that has been previously used, check that it has not been contaminated, that it is still fresh.

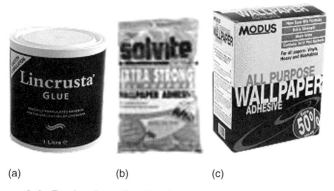

(a) (b) (c)

Figure 8.8 *Packaging of adhesives*

8.3
Pre-hanging information

Painters and decorators should prepare for the application of surface coverings by obtaining as much information as possible from manufacturers information. Prepare a checklist, here is an example:

- Read the manufacturers instruction leaflet supplied with each individual roll.
- Check visually that the roll is not damaged.
- If you require a number of rolls to paper a room check that the batch and shade numbers on all rolls are identical.
- *In situ* – open all rolls and check that they are not damaged by manufacturing printing or packaging.
- Identify the pattern. Is it a straight match or a drop pattern match?
- Determine which way up the paper should be applied. Some patterns are difficult to identify.
- If in doubt ask someone you are working with or consult the supplier or manufacturer.

Additional information that can be seen on the instruction leaflets.

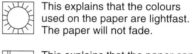

This explains that the colours used on the paper are lightfast. The paper will not fade.

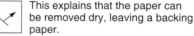

This explains that the paper can be removed dry, leaving a backing paper.

Figure 8.9 *Information pictograms*

Figure 8.10 *Drop match and Straight match*

8.3.1 *Patterns and economy of material*

The decorator will have to identify the pattern or texture match of a supplied surcace covering (wallpaper). There are a number of ways to establish this:

- First, look for the match on the manufacturers label in the form of a diagram.
- Second, open the roll, select a recognisable part of the design on the edge of the roll and measure the width of the paper, usually 530 mm from that point down the edge. At the 530 mm mark the same part of the design should be recognisable. This is the repeat pattern. This method is useful if the paper design is a texture and shapes are not easily recognised.

If the pattern is identified as a straight match, your lengths can be cut from one roll at a time, but if the pattern is a drop match, your lengths should be cut from two rolls as follows:

- Place two rolls on the paste bench.
- Match the pattern at the edge using the two rolls.
- Trim both top edges so the waste is equal.
- Name one roll 'A' in pencil on the reverse.
- Name one roll 'B' in pencil on the reverse.
- Cut lengths alternately from roll 'A' and 'B'.
- Keep the lengths in order when pasting and hanging.

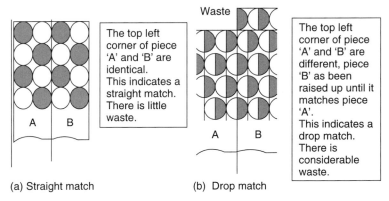

(a) Straight match   (b) Drop match

Figure 8.11 *Pattern matching*

Starting positions
Open the selected roll of wallpaper, look at the pattern or texture design on the paper and determine the scale of the pattern repeat. The standard dimensions of a roll of wallpaper is 10.05 m × 0.530 m. The width of a roll of wallpaper is usually the full design pattern repeat. A straight match can be found every 530 mm and a drop pattern match can be found 265 mm from the start of the design repeat. When you have identified where the pattern repeat will occur you can decide the starting point of the repeat for cutting and hanging purposes. All rooms are not identical in measurement therefore it is necessary to seek some advice regarding issues that could affect the finished result effect of the applied paper.

If patterned paper has been selected for use in a room, a balanced effect is visually required. It will be necessary to centre the paper on the main feature wall and work in two directions.

Balancing
To enable best effect of wallpaper to surfaces the design repeat motif should be measured and then divided into the height and width of a wall to determine starting points on feature walls such as chimney breasts.

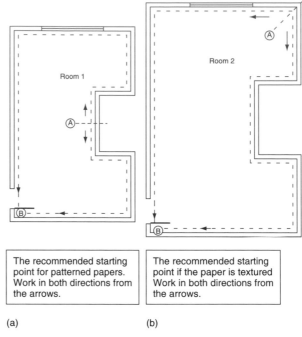

The recommended starting point for patterned papers. Work in both directions from the arrows.

The recommended starting point if the paper is textured Work in both directions from the arrows.

(a)   (b)

Figure 8.12 *Starting points in rooms*

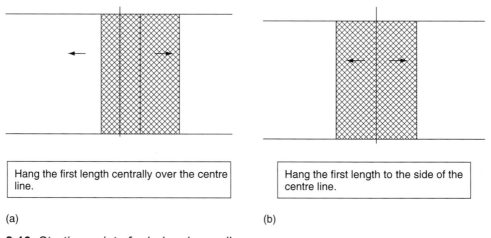

Hang the first length centrally over the centre line.

Hang the first length to the side of the centre line.

(a)   (b)

Figure 8.13 *Starting points for balancing wallpaper*

8.3.2 Staircases

All living spaces are different, the dimensions vary, the windows and doors are different in each individual room, there are corners, niches arches and other features. Planning of working procedure for the paperhanging activity prior to commencing the work is good practice to reduce the problems that could be encountered. Start at a light source, near a window and work towards the main user entrance.

Staircases are the exception

The paperhanging of staircases requires a different working strategy. The work requires planning of activities in two directions to enable the whole application procedure to match on completion and to reduce the waste that could occur through lack of planning. All staircases are different in layout and height, and the following list is a recommendation only:

- Plumb a vertical guide line in the stairwell on the longest length and hang the first length of wallpaper to this guide line (use concertina folds).
- Work from this first length upwards to the landing.
- Next hang the facing wall in the well.
- Continue plumbing on to the opposite well wall and paper back to the top of the landing.
- Complete the upstairs paperhanging, remember to reverse the butt joins when working away from the light.
- Try to finish upstairs in a corner that is not visibly conspicuous.
- Continue paperhanging down the stairs to the hallway.
- Complete papering the hallway, again taking into consideration the light source.

Accuracy is required with measuring, cutting and application at all times.

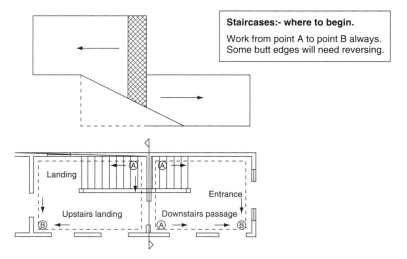

Figure 8.14 *Staircase working direction indicators*

8.3.3 Application of coverings to ceilings

The application of surface coverings to ceilings is not as difficult as perceived; consider that there are not as many obstacles to trim around. The only difficulty is managing the transfer of the paper from the bench to the ceiling bed. This requires the setting up of a suitable access scaffold to allow the paperhanger to walk the length of the room to enable transfer of the ceiling paper to the ceiling as the paper is unfolded from its concertina.

Due to the fact that most rooms have more than one entrance the starting place on each individual ceiling will have to be considered. Take into account the

- light source
- entrance doors
- length and width dimensions of room
- economy of material.

8.3.4 Recommended working procedure

- Measure and strike a chalk line centrally on the ceiling bed.
- Apply the paper to this line and work both ways from the line to the opposite walls.
- Trim each length after it has been applied, never trim as you go, adjustments to the paper might have to be made.
- Reverse joins as required to prevent shadows forming and detracting from the quality of the work.
- No overlaps of butt joins are permissible, or are gaps between the joins.

If the ceiling paper has no pattern and is a random texture, a chalk line can be struck from the selected starting point of the wall and work away from the light.

Ceiling starting points

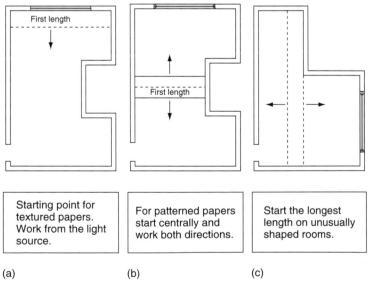

Starting point for textured papers. Work from the light source.	For patterned papers start centrally and work both directions.	Start the longest length on unusually shaped rooms.

(a)   (b)   (c)

Figure 8.15 *Ceiling starting positions*

8.4
Pasting and folding

The pasting and folding of wallpaper prior to application is one of the most important activities to take place. The outcome of the quality of the completed work depends upon such actions.

- Set up the paste bench and assemble the required measuring, trimming and cutting tools.
- Mix the adhesive and place in the correct location next to the bench ready for use, protect the floor surface.
- Have the wallpaper stored or placed ready for use, either racked under the bench or in the delivery container next to the bench.

The tools and equipment required for the pasting and folding process are as follows:

- paste bench
- adhesive container and agitator
- adhesive applicator
- measuring implements such as rulers, tapes
- paperhangers shears
- trimming knives and straightedge
- waste container
- protective coverings
- cloths and sponges.

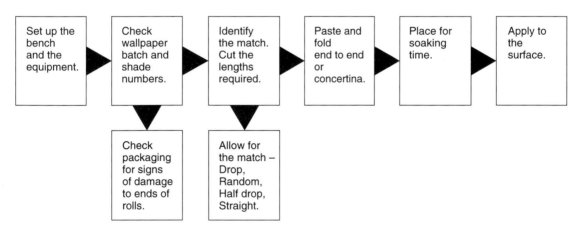

Figure 8.16 *The bench to surface process of paperhanging*

8.4.1 Recommended working practice for paperhanging procedures

Only one person measures, cuts lengths of wallpaper from the roll and applies the adhesive and then folds ready for application to surfaces.

- Lay a clean dustsheet on the floor in the work area.
- Erect the paste bench.

- Place a newspaper under the paste bucket to the right of the bench, work from right to left.
- Never place the paste bucket on the bench.
- Never allow adhesive to contaminate the bench.

Equipment set up

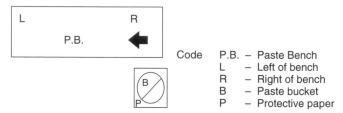

Figure 8.17 *Paste bench set up*

Lap or end-to-end folds
The lap or end-to-end fold is the accepted method of pasting and folding lengths of wallpaper that is to be applied vertically to walls. No paste should be allowed to contaminate the bench or the face side of the wallpaper. The top fold of the wallpaper should be two-thirds of the length when folded and the bottom of the wallpaper length is one-third. See illustration below.

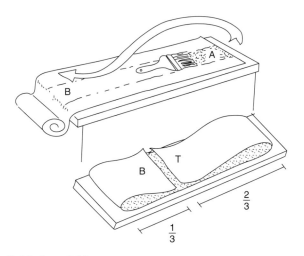

Figure 8.18 *Lap fold*

Concertina fold
The concertina fold is used when wallpapers are to be hung horizontally to walls and ceilings or for extra long lengths such as on staircases or foyers.

Care should be taken when folding to ensure that the edges meet thus reducing the risk of paste contamination. When applying the adhesive

- paste a bench length
- leave the first 50 mm of the paper unpasted
- make 300 mm folds
- continue with this procedure until the full length is pasted and folded
- use a support (short end of a roll) to move the paper to the soaking area or the surface for application.

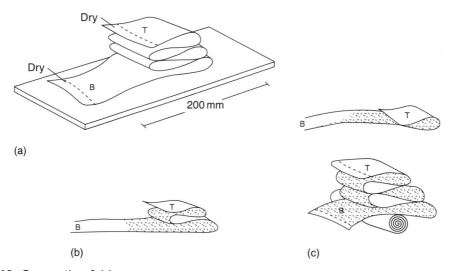

(a)

(b)

(c)

Figure 8.19 *Concertina fold*

8.4.2 Pasting procedure

The application of adhesive to wallpaper is a process that should be followed rigidly. Far too often it can be seen that poor practice takes place and the paste bench and the face of the wallpaper is contaminated. Follow these recommended stages and the application of the paper will become a clean process.

- Stand in front of the paste bench and place the cut length on the bench, reverse side upwards.
- Align the edge of the paper with the furthest edge of the bench and have a slight overlap to the short right edge of the bench.
- Apply adhesive down the centre of the cut length.
- Continue to apply adhesive from centre to edge, never edge to centre otherwise adhesive will creep under the edge and onto the face of the paper.

- Reposition the cut length to the nearside edge of the bench and apply adhesive from centre to nearside edge, never the reverse.
- The cut length is now ready to receive the top folds, either lap or concertina. Move the folds to the top right of the bench and continue the application of adhesive as before.
- Complete the folds and place the cut piece for soaking or application.
- Continue with the next piece.

Pasting and folding procedure for vertical lengths.

1 Place the paper (at length) on the paste-bench and slightly overlapping the farther edge.

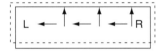

2 Place along the centre of the paper from right to left and then to the farther edge, from right to left.

Pasting and folding procedure for horizontal application.

1 Use the same pasting procedure as for vertical application.

2 When folding, make small pleated folds.

200 mm

3 Move the paper to the nearer edge of the paste-bench and paste from right to left, from centre to the near side edge.

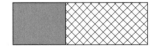

3 Leave the first and last 25 mm of each length unpasted 50 mm to stop paste creeping onto the face of the paper.

4 The paper should now be ready for transfer from bench to wall. When handling, use a short roll end to support the pleated folds.

4 Now make the first fold as shown (top fold $\frac{2}{3}$).

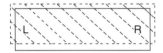

5 Paste the remainder of the length and make the bottom fold ($\frac{1}{3}$).

(a)   (b)

Figure 8.20 *Pasting procedure*

Pasting procedure for ready pasted papers (vinyls)

Some of the modern vinyl wallpapers have adhesive applied during manufacture. This is in the form of dry adhesive crystals. All that is required when this type of paper is to be hung is to immerse the cut length into a trough of water and apply the paper immediately to the surface from the trough. Many traditional decorators prefer to mix a thin solution of adhesive. Apply this adhesive to the pre-pasted paper. Fold in the conventional manner and then apply the paper to the surface. A stronger bond with the surface is achieved through this double pasting.

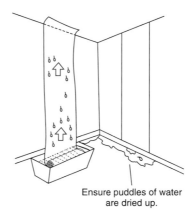

Ensure puddles of water are dried up.

Figure 8.21 *Trough method of application*

Application of adhesive to borders

When adhesive has to be applied to borders cover the paste bench with lining paper. This will prevent contamination of the bench as the strips are pasted and folded. Two to three widths can be pasted on the bench depending upon the width of the border. Replace the lining paper as required.

Good working practice checklist

- Have a clean towel, bucket, water and flannel handy to keep your hands free of adhesive contamination during the paper hanging activity.
- Have a waste receptacle handy to place all cuttings in as you work.
- Clean the edges of the paste bench frequently to remove any adhesive that may have collected.
- Clean shears frequently during use to keep them sharp.
- Clean all paper hanging equipment at the end of a working shift.
- Check surrounding surfaces for adhesive contamination. Adhesive can damage polished furniture and floors.
- If delays are encountered during the application process open up any pasted lengths to dry. They can be repasted later.

Do's during the pasting and folding process

- Remove any lumps from mixed adhesive.
- Remove hairs from pasted lengths.
- Leave the first and last 50 mm of each pasted length for ease of handling.
- Keep edges parallel when folding.
- Work from right to left on the bench.
- Keep all tools and equipment clean.

Don'ts during the pasting and folding process

- Leave unpasted edges or areas on the paper.
- Contaminate the face of the paper or the bench.
- Mix the paste too thinly.
- Place the paste bucket on the paste bench.
- Fold the paper incorrectly.
- Allow other persons to assist with the pasting and folding process.
- Leave tools and equipment in a dirty state.
- Have no floor protection under and around the paste bench.

8.5 Trimming procedures

The trimming of wallpapers requires skill to ensure that visible edges at and around obstacles are accurate. Nothing detracts from the finished appearance of applied wallpaper than uneven trimming.

Most manufactured wallpapers come with the edges pre-trimmed during manufacture so the joining of lengths is no great problem. Expensive wall coverings however are individually boxed and have a protective strip along the length edge of each roll. This protective edge is called a 'selvedge'.

Prior to hanging these papers the selvedge must be removed by the use of one of the following methods:

- a Ridgley straight edge and trimmer
- a metal straight edge and trimming knife
- Shears.

The process of removing the selvedge must be accurate otherwise the edges will not meet.

8.5.1 Ridgley straight edge and trimmer

Consists of a stainless steel machined straight edge one metre in length that holds a cutting wheel in place along one edge of the track. When this wheel is moved along the track there is no variation in movement thus allowing an accurate cut. A protective zinc strip is incorporated in the underside of the track and this is removed and placed under the paper to be trimmed to protect the cutting wheel from becoming blunt or damaging the bench surface.

How to use:

- Place the zinc strip on the paste bench under the paper where the cut is to be placed.
- Place the straight edge in position on the paper where the cut is to be made.
- Place the cutting wheel in the running track.
- Hold the straight edge in place, firmly (two persons) press the wheel handle down to engage the cutting blade and progress along the length of the track.
- A precision cut should be obtained.

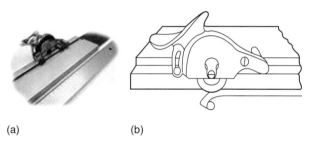

(a) (b)

Figure 8.22 *Ridgley straight edge and trimmer*

8.5.2 Metal straightedge and trimming knife

A variety of lengths of straight edge can be used to trim wallpapers. It is a versatile method used for the trimming of edges, borders or mitres. The straight edge and knife can be used on the bench or on the wall.

Very fine pencil marks are dotted on the wallpaper to indicate where a cut is required. The metal edge is placed on these marks and the cut is made with the knife. Use new blades and change the blades frequently to ensure the best cuts are achieved.

Figure 8.23 *Metal straight edge and trimming knife*

8.5.3 Paperhanger's shears

A hand process of trimming and requires operator skill in managing an accurate cut around obstacles.

Trimming procedures include the following:

- at ceiling wall and skirting angles
- around door and window casings reveals, frames or architraves
- around electrical switch plates and sockets
- around and behind radiators
- around fireplaces, niches and arches
- at internal and external corners
- trimming on the bench.

All require the use of sharp tools to achieve a quality trim and various techniques are adopted to enable the trimming process to be carried out effectively. When applying marks to wallpaper to enable trimming never use a ball pen as the ink can bleed through the vinyl. A 2H pencil is the best marking implement or the tip of the shears.

Procedures for trimming at ceiling angles and skirting

- Apply the length in position on the surface using the plumb line guide marks.
- The waste at the top and bottom of the length should be tucked into the angles with the paperhanging brush or caulker.
- Use the shears to mark along these edges.
- Pull the paper away from the ceiling and trim with the shears, make the cut along the outside edge of the crease mark.
- Repeat the process at the skirting line.
- Tuck the trimmed paper back to the surface with the paperhangers brush.
- Continue the process with the next length.

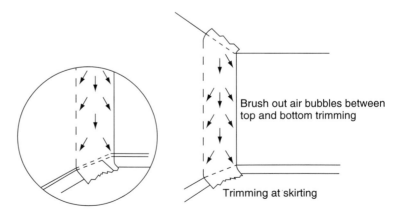

Brush out air bubbles between top and bottom trimming

Trimming at skirting

Figure 8.24 *Trimming at angles*

Trimming at internal corners

Not all corners are plumb or vertical, it is necessary to allow some adjustment to be made to the wallpaper when turning corners. A slight overlap of the paper is required to account for the corner being out of plumb.

- Measure the distance left from the last hung length to the corner, measure at the top middle and bottom of the wall.
- Select the longest measurement and allow 5–10 millimetres for the overlap around the corner.
- Trim the strip required on the paste bench.
- Apply the measured strip to the wall and tuck the overlap around the corner, then trim.
- Apply the remaining strip and plumb to vertical.
- Commence hanging strips along the wall.

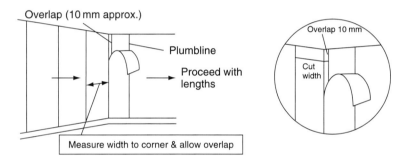

Figure 8.25 *Internal corners*

Trimming procedure at reveals

The application of wallpaper on and around reveals requires some planning to minimise the effect of required overlaps or secret joining. Wallpaper is required to cover two places but once the trim is completed only one area can be covered. A second piece is then required to complete the process and the application of this second piece has to be accurate. As paper is hung over and around the reveal, plumbing of the lengths becomes important to keep the paper and pattern running true.

- Position and hang the first full length.
- Create accurate cuts horizontally at top and cill reveal.
- Smooth out, check for plumb and trim at ceiling and skirting.
- Apply piece 'B' and trim at ceiling and window edge.
- Apply piece 'C' and trim at cill and skirting.
- Cut and apply piece 'D' with an overlap and cut through both layers of wallpaper with a new blade, remove the front and rear waste.
- An invisible join should have been created.

- Continue applying pieces of wallpaper and adopt the same procedure as you leave the window reveal.
- Plumb and continue paper hanging.

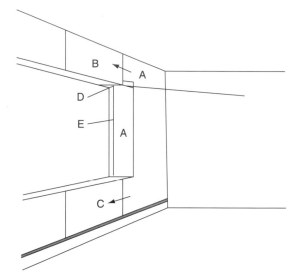

- apply matching piece of paper
- overlap by 50 mm
- trim through both layers with a sharp knife
- remove waste and smooth down the paper to make an invisible join.

Figure 8.26 *Trimming at reveals*

Trimming procedures at and around sockets and switches
To remove wallpaper where switches and sockets are present the following procedures should be followed:
For circular plates

- turn off the electricity supply
- paper over the plate
- push the shear point through the paper at the centre of the plate
- from the centre create lots of small cuts to the circumference edge of the switch (called a star cut)
- tuck the paper flat and trim off the waste with a sharp knife
- clean off any adhesive.

For rectangular or square plates

- switch off the power supply
- loosen switch plates with the correct screwdrivers
- apply paper over the plate and push the point of the shears through the centre position of the plate
- make a sharp cut to each corner of the plate
- allow slight waste to tuck behind the plate and trim off the remaining paper
- replace the switch plate and clean off any adhesive from the plate
- switch on the electricity supply.

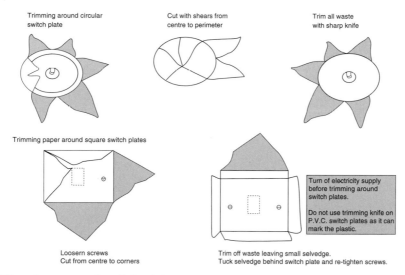

Figure 8.27 *Trimming around switch plates*

Trimming around fireplaces
To ensure wallpaper is trimmed correctly around fireplaces and mantles, make a series of small cuts from the overlap waste of the wallpaper to the moulding outside edge. Patience is required but the results are rewarding.

- hang the length over the fireplace
- brush and smooth out the paper down to the mantelpiece
- make a series of small cuts as shown on the diagram
- tuck in the paper to the wall surface
- trim off the waste using a sharp knife.

Trimming around radiators
If it is not possible to remove or drop the radiator it is still possible to paper around such obstacles.

- Apply the paper to the wall and smooth out to the top of the radiator and the edges of the two mounting brackets.
- Make a pencil mark where the brackets touch the paper and ease the paper back from the wall.
- Make a vertical cut from the bottom of the wallpaper up to the mark.
- Tuck the paper down behind the radiator and mark the bracket shape on the wall paper.
- Remove the paper again and trim the bracket shape off the paper.
- Tuck the paper behind the radiator and join at the bottom edge near the skirting.

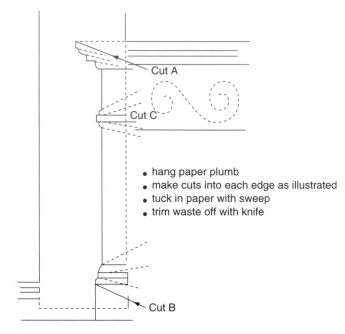

* hang paper plumb
* make cuts into each edge as illustrated
* tuck in paper with sweep
* trim waste off with knife

Figure 8.28 *Trimming around fireplaces*

* Use a radiator roller to smooth down the paper to remove air bubbles.
* Trim at the skirting edge.
* Clean off any adhesive from the radiator and surrounding areas.

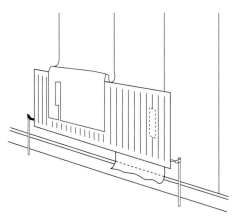

Figure 8.29 *Trimming around radiators*

9 Colour in decoration

9.1 Introduction

Colour is an integral part of painter's and decorator's profession. Knowledge of colour, what it is and how it can be used is required to enable advice and guidance to be offered to customers. This advice when offered will allow customers to put together personal room schemes, individual to their own preferences.

The painter and decorator should have knowledge of the following aspects of colour such as:

- What is the spectrum?
- Can you name the colours of the spectrum in the correct natural order?
- What is white light?
- What are primary colours?
- What are secondary colours?
- What are tertiary colours?
- What are light primaries?
- What are pigment primaries?

Knowledge of colour terms is required to enable explanation of schemes to clients. The basic colour terms are as follows:

- achromatic colours
- complimentary colour
- monochromatic colours
- contrasting colours
- discordant colour
- split complementary
- tints and shades.

It is not a simple or easy thing to understand colour, we just take it for granted, but if used correctly it can significantly enhance our living environment.

9.2
Theory of colour

Light in its natural form is transparent and you cannot touch it. To understand colour we call light 'white light'. Light travels from the sun to the earth, the shortest distance it has to travel is noon and if there is no atmospheric pollution the sky would be seen as blue. The light waves are termed short waves, thus short wave lengths are blue.

Figure 9.1 *Short wavelength*

At dusk and dawn the light travelling from the sun hits the earth on a tangent and thus travels further. These light rays are coloured red; they are termed long wave lengths.

Figure 9.2 *Long wavelength*

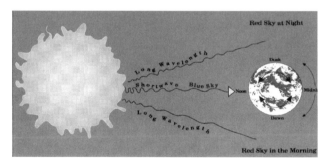

Figure 9.3 *Light waves from the sun to the earth (see Plate 1)*

9.2.1 The spectrum

When white light is broken up into its parts by the use of prisms many variations of colours can be seen. The main colours however are:

- red
- orange
- yellow
- green
- blue
- indigo
- violet.

Definite divisions between the colours cannot be defined. Isaac Newton, through research and experimentation, found that if he passed one ray of white light through a prism then that ray of light would emerge from the other side in all its parts. It emerged as a spectrum containing all the colours of the rainbow, each merging into each other.

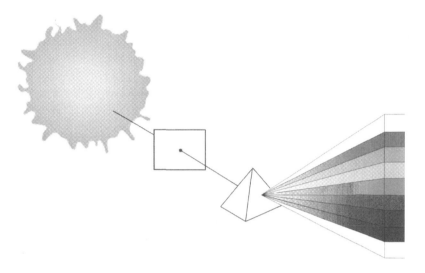

Figure 9.4 *Creation of the spectrum (see Plate 2)*

9.2.2 How we see colour

The natural order   To make sense of colour, systems had to be devised that would allow understanding and put colour into some form. A system that has been used successfully is the natural order or the colour circle.

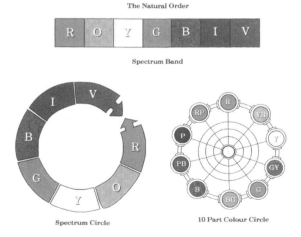

Figure 9.5 *The natural order and colour wheel (see Plate 3)*

Activity 9.1

Using the following materials

- drawing paper/cartridge quality or watercolor
- 2H pencil
- compass
- ruler
- suitable medium of watercolours/postercolours or gouache
- artist brushes.

construct a colour circle with six sections, apply the primary colours, apply the secondary colours, construct two more colour circles with six sections.

1. produce a tint circle of colour by adding white to the primary and secondary colours
2. produce a shade circle of colour by adding black to the primary and secondary colours.

Include your work as evidence in your portfolio.

Activity 9.2

Colour activity exercise
Draw a rectangle 11 cm in length and 1 cm wide, and divide into 11 sections, then

- paint one end square in white
- the opposite end square in black
- the middle square in a neutral grey.

The remaining squares should be filled in with greys of varying value. Your completed activity should produce a value scale from black to white.

Pigment primary, secondary and tertiary colours
Colour used in the industry is not the same as colour used by theatre and stage. We use coloured pigments that are obtained from the earth or manufactured. They are known as natural pigments or lakes and dyestuffs.

- Pigment primary colours cannot be produced by mixing other colours together, they are pure colours.

Figure 9.6 *Pigment primary colours (see Plate 4)*

- Pigment secondary colours are produced by mixing two primary colours together.

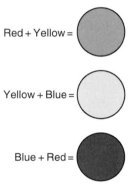

Red + Yellow =

Yellow + Blue =

Blue + Red =

Figure 9.7 *Secondary colours mixed from primary colours (see Plate 5)*

- Pigment tertiary colours are produced by mixing two secondary colours together.

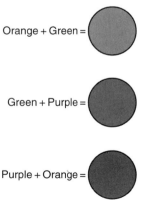

Orange + Green =

Green + Purple =

Purple + Orange =

Figure 9.8 *Tertiary colours mixed from secondary colours (see Plate 6)*

The technical term used to describe the intermixing of pigment colours together is called 'subtractive admixture'.

Light primary colours

These are the parts of white light that have no tactile body. They are rays of light. The three light primaries are red, blue and green. When these lights are intermixed they form white light. The technical term is called 'additive admixture'.

9.3 Terminology of colour

To put some meaning into colour terms and descriptions the colour wheel is used as an aid. The following descriptive terms can then be adapted to suit circumstance. The colour consultant can manipulate this understanding of colour to offer a service that is unique and individual to a client's requirement.

- *Achromatic colours* – These are a range of tones from black through to white. Black and white are known as sensations, not colours.
- *Advancing colours* – These are warm colours taken from the red, orange and yellow section of the colour circle. They give the sensation of coming towards the eye.
- *Analogous colours* – Two to three colours next to each other in the colour wheel or the natural order, complemented by a contrasting colour.
- *Accent* – Colours selected that lift colour schemes, usually a contrasting colour in small quantities placed strategically within the colour scheme.
- *Contrasting colours* – These are opposite each other on the colour wheel, green is opposite red. Blue is opposite orange. Purple is opposite yellow.
- *Discordant colour* – This is where the natural order has been reversed. The stronger colour becomes the weaker and is not visually pleasing to the eye.
- *Harmonious colours* – If three or more colours are used together in a scheme and the general effect is pleasing to the eye, harmony has been created.
- *Juxtaposition* – The effect one colour has over another when placed side by side. The complementary of each colour can distort the appearance of the actual colour present giving the effect of more colours being present.
- *Monochromatic colours* – Tints and shades of one colour by adding black or white to create a tonal, tint and shade scale.
- *Natural order of colour* – The lightest part of the spectrum is yellow and the darkest purple, in this order they are pleasing to the eye. If the darkest colours became the lightest, they would be unpleasing to the eye (discordant).
- *Split complementary* – On colour circles with upwards of ten sections the true opposite contrasting or complementary colour will be a combination of two colours. Refer to the 10-part colour wheel where the split complementary of red is blue/green.

- *Tint* – The addition of white to a hue.
- *Shade* – The addition of black to a hue.

9.4
BS 4800 paint colours for building purposes

To put some purpose to colour theory, systems had to be devised that would allow manufacturers to produce the product and for the decorator to use the product.

The BS 4800 system in use today has evolved from other systems which were:

- A dictionary of colour for interior decoration produced by the British Colour Council.
- BS 381 C colours for specific purposes.
- BS 2660 colours for building and decorative paints.

The BS 4800 framework:

- There are 5 sections – A, B, C, D, E – and these are known as the weight sections from 'A' the weakest in colour saturation to 'E' full strength colour saturation.
- There are horizontal rows of colours from 02 red/purple through to 24 purple and an extra row coded 00 for neutrals such as black and white. These are known as the hue rows.
- There are vertical rows from 01 to 55 to indicate the greyness of a colour. 01 is in Section 'A', practically neutral colours to 55 in Section 'E' which has no grey content.

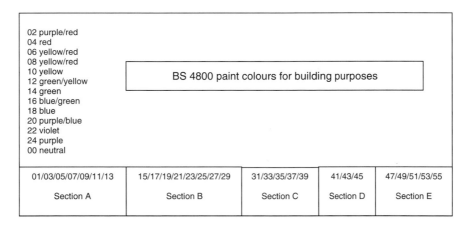

Figure 9.9 *BS layout of framework*

The BS 4800 system was developed to

- only produce enough colours for general use
- meet design and technical requirements
- build a relationship between colour and building materials
- meet economic trends.

<table>
<tr><td>**Key point**</td></tr>
<tr><td>The Munsell system of colour notation has influenced the development of colour systems in use in the UK. It is worth some research to fully understand the evolvement of colour order and coding.</td></tr>
</table>

The framework consisted of 100 colours including black and white, there was a

- Basic selection of 30 colours plus black and white produced by all manufacturers.
- Supplementary selection of 68 colours, these colours will not be available from some manufacturers due to production costs. There are two sub sections named 'G' and 'M'. Sub section 'G' covers oil based coatings in full gloss or eggshell and there are 45 colours available. Sub section 'M' covers emulsions, matt and silk and there are 34 colours available.

To identify these colours on manufacturers colour cards, symbols are used as follows:

Full gloss finish in oil

Semi gloss finish

Matt finish in emulsion

Silk finish in emulsion

Figure 9.10 *Identification symbols*

Colour codes created from BS 4800 paint colours for building purposes and placed on paint containers to identify the colour content are in three parts:

- The first part is an even number and indicates the hue.
- The second part of the code is a letter and this indicates the greyness of the colour.
- The third and final part of the code is usually a odd number and indicates the weight of the colour.

Thus a typical code would be 08 B 15 which is a colour known in the trade as magnolia.

The colour clock – This is a teaching aid to help users of the BS 4800 system. A technically competent painter and decorator will through time become familiar with the BS 4800 system. A colour code written on a job schedule or specification form will be instantly recognisable in the mind of the painter. The use of a colour card helps reinforce that mental picture.

Look at the colour circle and imagine it as the face of a clock. There are 12 hue codes from 00 to 24.

1. 02 is at 1 o'clock
2. 06 is at 3 o'clock
3. 12 is at 6 o'clock
4. 24 is at 12 o'clock.

Think of the hour and multiply by 2, this will give you the hue.

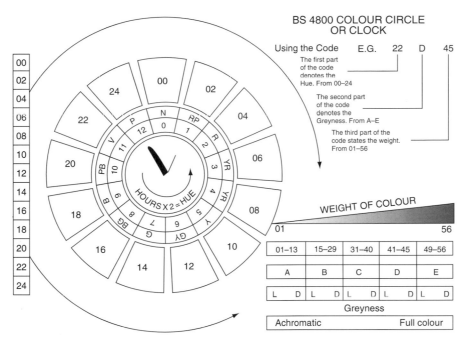

Figure 9.11

Start with the red–purple and finish with purple. Work through the hue codes as you would through the natural order. Thus red/purple, red, yellow/red, yellow/red, yellow, green/yellow, green, blue/green, blue, purple/blue, violet, purple.

Test your knowledge 9.1

1. Describe the term 'natural order of colour'.
2. What name is given to colours that cannot be mixed from other colours?
3. How are secondary and tertiary colours mixed?
4. How is white light broken up to show all the colours of the rainbow?
5. If black has been added to a colour it becomes a?
6. If white has been added to a colour it becomes a?
7. If black and white has been added to a colour what type of colour terminology is applied?
8. Describe the framework of BS 4800.
9. Name and describe the three parts of a colour code.
10. What free leaflet is available from the retailer for use in selecting colour that gives BS 4800 reference numbers?

Store your responses in your portfolio for evidence.

10 Decorative paint effects

10.1 Introduction

Decorative paint effects are created by manipulating paint mixtures to imitate natural materials such as wood and marble, and to create infinite textures by using a variety of tools brushes and applicators.

There are four categories of effect ranging from:

- Imitating natural woods by graining.
- Imitating naturally formed rocks to produce marble effects.
- Creating texture by using applied colour to surfaces. Broken colour is the term applied to producing textures.
- Stencil work.

In earlier years people could not afford to purchase natural materials to use in building construction work, thus the specialist painter was called in to re-create the materials. In the 1950s it was popular to have your outside woodwork grained to imitate hardwood. The use of texture paints in the 1960s and early 1970s required the technique of glazing and wiping to complete the effects created by stippling and combing of plastic compounds. Slate fireplaces were marbled to imitate the real product.

How times have changed, materials are readily available and at a cost that people can afford. It is the specialist painter and decorator that is the most expensive to employ to create specialist decorative effects and only the most affluent can afford these skills.

Restoration of historic buildings, national trusts seek out craftsmen who hold these skills and are able to transfer them where required in maintaining our national heritage.

10.2 Types of decorative effect

Graining   The imitation of a selected timber to a prepared surface using specially prepared coatings. Specialist tools and brushes are used to create the background grain texture and figure work. The two main timbers concentrated on at this level of work are:

1.   Oak – brush grained, dragged, combed, figured, quartered.
2.   Mahogany – brush grained, flogged, combed, figured, feathered, mottled.

Marbling   The imitation of pure marbles by applying specially pre-pared mixtures over pre-prepared surface grounds. It is produced by manipulating the applied mixtures with a selection of tools and brushes. It requires great skill and much practice before competence can be gained. The marbles covered at this level are:

- Carerra – a white marble quarried from Carerra in Italy.
- Vert de Mer – a black, green and white marble.
- Sienna – a buff , ochre and yellow veined marble.

Broken colour   A term used to describe the creation of texture to a coating or a base surface. The coating is applied to a surface as an additive, or removed from the surface as a subtractive texture. The types of decorative treatment that are included in this section are:

- sponging
- rag rolling, additive or subtractive
- hair stippling
- rubber stippling
- dragging
- combing
- bagging.

Stencil work   The application of coloured paints through specially prepared stencil plates to create applied design to surfaces. These can be as individual motifs, running borders or overall trellis work patterns.

Stencil plates can be purchased ready for use or prepared by the specialist decorator from any drawing available. The drawing can then be converted into a stencil, cut out, prepared and used. The types of stencil treatment covered at this level are:

- single plate stencils, positive or negative
- multi plate stencils
- edge stencils.

Gilding   This is the application of loose leaf or transfer gold or metal-lic foils to surfaces for enrichment. It is only the intention to briefly discuss this area and make you aware of the basic tools, brushes and materials used. It is a highly skilled practice.

10.3 Types of brushes tools and equipment

Figure 10.1 *Paint brush used for applying glazes and scumbles*

To create the many decorative treatments, there are numerous brushes used with a combination of tools and equipment. The saying goes 'if it can produce a texture pleasing to the eye, use it'.

10.3.1 Brushes

Laying in brush
Used for the application of base or ground coats to surfaces prior to application of the prepared glazes or scumbles.

Fitches
These are used to apply prepared mixtures to surfaces. The fitch can manipulate the applied mixture to create figure work or texture as required. There are a variety of sizes. They are available as flat, round or fan shape.

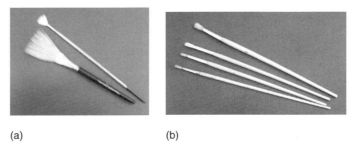

(a)   (b)

Figure 10.2 *Fitch styles*

Softeners
Produced in either hog hair of badger hair for use with water or oil based media. The badger softener is the more expensive. The function is to soften harsh edges left during the marbling or graining process. The softener blends the colours and shapes of the applied media to create realistic effects Various shapes are available, but mainly the flat or round softener.

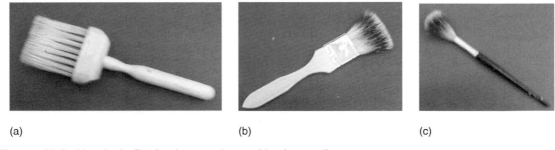

(a)   (b)   (c)

Figure 10.3 *Hog hair, flat badger and round badger softeners*

Figure 10.4 *Hog hair stipplers*

Hair stippler
A hog hair brush that is used to remove brush marks from applied oil based glazes or paints. A fine texture is produced which is decorative in its own right or over texturing can take place. Various sizes are available.

Flogger
A brush designed to remove brush graining marks and replace them with fine pore marks representing the background texture of most timbers. The bristle selection is of hog and horse hair mix.

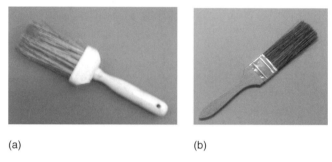

(a)   (b)

Figure 10.5 *Flogger*

Cutters, mottlers and overgrainers
The cutters are used to apply figure work to graining, as the name suggests the brush is pulled or pushed through the applied media to create the required figure effect. The mottler is used to create highlights in graining work. The overgrainer applies figure work as a second layer of effect to create depth and reality to the imitated work.

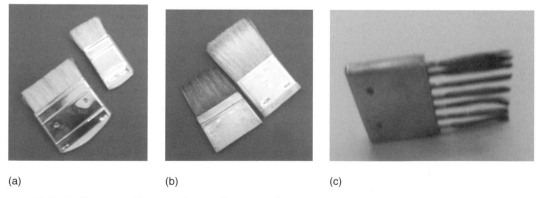

(a)   (b)   (c)

Figure 10.6 *Cutters, mottlers and pencil overgrainer*

Drag

This is a coarse bristled brush with a filling of fibre, nylon or horse hair. It is usually palm held and is pulled through graining colour to produce a coarse brush grain.

(a)   (b)

Figure 10.7 *Types of drag brush*

Figure 10.8 *Stencil brush*

Stencil brushes

These are of varying design and the bristle brushes are short in length for applying paints through stencil plates. If the bristles are full in length, the bristles and the paint loaded on them can creep under the stencil plates causing smudging.

Artist's selection of brushes

Consists of a sign writing pencil, one stroke and a variety of smaller brushes. These are used to apply detail to decorative effects such as veins in marble or figure detail in graining. They are also used for toughing up fine misses on stencil work or gilding.

Figure 10.9 *Artist selection*

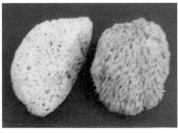

Figure 10.10 *Natural sponges*

Sponges

Both natural and synthetic can be used to create some pleasing effects in multi-colour. The natural sponge produces the best quality marks.

Figure 10.11 *Graining combs*

Figure 10.12 *Palette set up*

Graining combs
Constructed from tempered steel or rubber in a variety of widths. Fine medium and wide sizes are available, used to produce varying grain effects. Combination combs can be purchased.

Palette dipper and knife set up
Most of the mixing of colours for use with media will be carried out on the palette set up. Sample colours and glazes can be tried for suitability before mixing batches of material for use. Palettes can be purchased or made from hardboard or MDF. Attach a dipper which can hold oil, turpentine or gilp, and have to hand a small palette mixing knife. This set up will be adequate for most sample trials or small works.

Stencil knives
Implements for cutting stencil plates; these are extremely sharp knives and should be used with great care. Make all cuts away from the hands and the body to minimise the risk of cuts. Use with cutting mats to prevent damage to underlying surfaces.

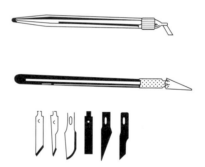

Figure 10.13 *Stencil knives and blades*

10.4 Types of materials

Materials used for creating decorative effects come under two categories:

1. oil based
2. water based.

The following is a general list of materials used to create the liquid media for application to surfaces.

- oil based undercoats and eggshell paints
- vinyl emulsions and acrylics
- transparent oil glaze
- acrylic scumble glaze
- oil scumbles
- linseed oil
- turpentine, substitute turpentine and white spirit
- methylated spirit
- tinters and stainers, oil, water or universal types
- terebine driers.

Other items that can be used to create texture range from good quality lint free cloth to all grades of paper or polythene. Old car inner tubes make excellent rubber stipplers. Washleathers produce the best crushed velvet rag rolling effects.

Base or Ground coats The surface to which specially prepared media is applied and then worked into a texture should be either water or oil based. The best oil based surface is eggshell as the brush marks can be wet abraded from the surface leaving a plastic like, totally flat surface ready to receive the prepared medium. This medium can then be textured.

Oil based mediums do not work as well on water based base or ground coats. Water based scumbles and media should only be used on emulsion or acrylic base or ground coats. Specialist effects can then be applied to these base or ground coats in one of two processes:

- in layers built up as each previous layer dries to create depth and detail
- by manipulating applied glaze and adding or removing media until the desired effect is achieved in one application.

Preparatory processes To achieve high standard base or ground coats treatments should be as follows:

> **Key point**
>
> - *Additive texturing* – the building up of a texture by applying coat after coat to the base.
> - *Subtractive texturing* – where media is applied over the surface and parts of the media is removed using tools, brushes or other material.

<table>
<tr><td>

Untreated surfaces
Oil based Grounds

- dust off, abrade dry and re-dust
- apply primer or sealer to the surface
- face fill, allow to dry, sand level and apply undercoat
- dry sand and apply second undercoat
- wet abrade to remove brush marks
- apply first eggshell
- wet abrade
- apply second eggshell
- wet abrade and dry off the surface

The initial decorative treatment can now be applied. On completion of specialist effect apply a protective clear coating.

</td><td>

Untreated surfaces
Water based Grounds

- dust off
- abrade dry and re-dust
- apply primer or sealer to the surface
- face fill, allow to dry sand level and apply first acrylic or emulsion coat
- dry sand and apply second acrylic or emulsion coat
- apply final acrylic or emulsion

The initial decorative treatment can now be applied. On completion of specialist effect apply a protective clear coating.

</td></tr>
</table>

Figure 10.14 *Preparation processes*

The preparation of the base coat is of extreme importance as the final applied decorative effect will not be achievable without a well prepared flat surface being available to create the technique.

Activity 10.1

Prepare a number of surfaces to enable the practicing of the decorative techniques in your own leisure time. Painted hardboard, MDF, or flat vinyl wallpaper with eggshells and vinyl emulsions will provide base coats for applying mixed media. An adequate and manageable size is A4 or A3. Mixed varnish and coloured gloss if thinned to the correct consistency will provide a suitable media with which to practice decorative effects.

10.5 Descriptions and processes

10.5.1 Sponging

The application of colour using a natural or synthetic sponge. Colour is loaded on to the sponge and this colour is then transferred to the ground or base coat. The colour is tipped to leave fleck shaped marks on the surface. A variety of shapes can be created. When dry, other colours can be added leaving a random all over multi-fleck effect. Emulsions are the best media to work with.

Equipment/paint

- roller tray to hold paint supply
- natural or synthetic sponge
- oil or emulsion paint
- lining paper for test sample
- base colour and brush
- cleaning materials
- various colours.

Note:
Clean out all equipment as required. Remember natural sponges are expensive items to purchase.

Procedure
Natural sponges produce the best broken colour effects.

- Pour the selected colour on to the roller tray.
- Moisten the sponge with the appropriate solvent.
- Dip the sponge into the colour.
- Squeeze out excess colour.
- Produce a sample print.
- Proceed to sponge the surface in a random application manner.
- When dry apply second colour.
- Three colours are very effective.

10.5.2 Stippling

The effect where the applied medium is worked using a selection of brushes to create an even texture which shows no application brush marks. The overall finish can be fine or coarse. Hair or rubber stippling are best carried out in oil based media. Other effects such as blending or shading can be added to the effect. After application of oil glaze to the surface by brush and prior to the hair stippling itself the tips of the stippling brush are finely coated with glaze from the application brush to prevent patchiness occurring.

Equipment/paint	Procedure for oil glaze stippling
<ul><li>roller tray and paint kettle for the mixed glaze</li><li>turpentine</li><li>driers</li><li>tinters/stainers</li><li>cloths</li><li>paint brushes or rollers for application</li><li>hair or rubber stipplers</li><li>sponges.</li></ul>	<ul><li>Mix up the glaze, add colouring, a small amount of turpentine and a drop of driers.</li><li>Apply the mixed glaze to the surface.</li><li>Even out the brushmarks with the hair stippler.</li><li>Stipple in a random pattern to obtain an even texture.</li><li>Clean appliances on completion.</li></ul> Procedure for water paint stippling <ul><li>Mix up the scumble glaze or paint.</li><li>Dip the applicator into the mixture.</li><li>Squeeze out excess.</li><li>Apply the applicator to the surface in a random pattern.</li><li>When dry add contrasting colours.</li><li>Clean appliances on completion.</li></ul>

10.5.3 Rag rolling

An effect created by rolling a crumpled piece of paper, rag or wash leather through applied glaze. Part of the applied glaze is removed leaving a crushed velvet effect. This is called subtractive texturing.

Equipment/paint	Procedure
<ul><li>mixed and coloured glaze</li><li>applicator brush</li><li>a hair stippler to create an even background texture</li><li>kettles or roller trays to hold glaze mixture</li><li>an adequate supply of application cloth.</li></ul> Note: Use one type of cloth otherwise the texture could alter.	Apply the coloured glaze to the ground and remove the brush marks with a hair stippler. Create a wad of application cloth and roll randomly over the surface. The cloth will lift off glaze and leave the desired texture. Note: <ul><li>Where ceiling and walls meet dab the rag to finish off the process.</li><li>Open out soiled cloths and allow them to dry before disposing of them otherwise spontaneous combustion could occur.</li></ul>

Additive rag rolling is where the cloth is soaked in medium and the cloth is rolled randomly over the base or ground coat to produce the effect. Two or more colours can be added.

10.5.4 Dragging and combing

An effect created by pulling brush bristles or combs through applied glaze. A series of lines are created with some texture to the lines. The manipulation of combs through the media can create a variety of patterns. Applicator skill is required when using combs.

Equipment/paint	Procedure
• oil glaze and stain • paint kettle • brush for application • rags for cleaning • drag brush • metal/rubber combs • cardboard. Some of the specialist brushes and combs are expensive. Home made varieties can be produced from cardboard.	• Apply the glaze sparingly to the surface and lay off in one direction. • Use the drag to create uneven lines in the glaze. The idea is to let the ground coat show through and create a two tone effect. • Use the combs at all angles to produce various effects. • Patterns such as herringbone and basket weave will require practice.

Broken colour samples

(a)

(b)

Figure 10.15 *Sponge stippling (see Plate 7)*

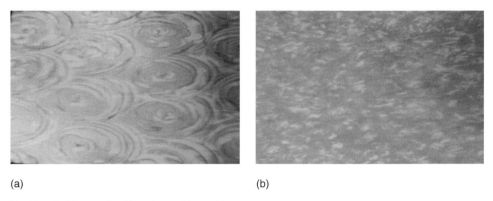

(a)   (b)

Figure 10.16 *Rubber stippling (see Plate 8)*

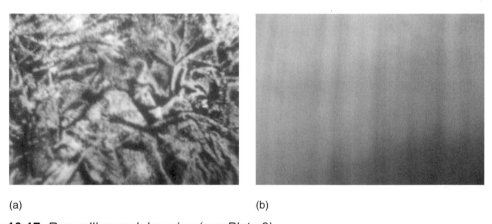

(a)   (b)

Figure 10.17 *Rag rolling and dragging (see Plate 9)*

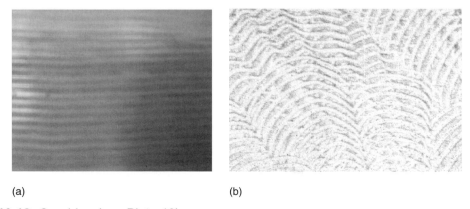

(a)   (b)

Figure 10.18 *Combing (see Plate 10)*

10.5.6 Marbling

Carerra marble

The purest Carerra marble is pure white with no visible veining. It is a semi-translucent stone with lustre. Grey veins appear in the less pure product and the ground is usually mottled. Ground oxides are also present which can offer blue, green or yellow tonal discolouration to the marble.

Tools	Method/technique
<ul><li>lint free cloth</li><li>palette board</li><li>fitch</li><li>hog or badger softener</li><li>artist brushes</li><li>sign writing pencil</li><li>paint kettle.</li></ul>**Materials**<ul><li>white eggshell in oil</li><li>payne grey oil stain</li><li>yellow ochre</li><li>raw umber</li><li>black</li><li>ultramarine</li><li>transparent oil glaze</li><li>linseed oil</li><li>turpentine</li><li>varnish.</li></ul>	<ul><li>Wipe gilp over the surface of a ground prepared in white eggshell oil.</li><li>Mix three transparent glazes in tones of Payne's grey and apply to the panel in a cloud formation, obtaining a non-distinguishable texture.</li><li>Soften the whole panel with the hog and badger softeners in all directions.</li><li>Apply tonally deeper grey in angular lines and soften from one edge only.</li><li>Mix a deeper grey apply and soften again, add fine veining to obtain details.</li><li>Mix a glaze plus thinned white enamel and apply over the created effect, this will produce a translucent effect.</li></ul>or<ul><li>Lay in the tinted glaze in a diagonal direction over the panel leaving areas of white ground.</li><li>Roll on tonal grey effect with a large fitch and soften from one edge only.</li><li>Add fine veining.</li><li>Varnish when completed.</li></ul>

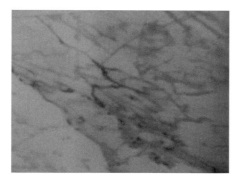

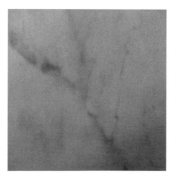

(a)   (b)

Figure 10.19 *Carerra marble examples (see Plate 11)*

Vert de Mer

This is a black and green marble with white veining. There are a variety of procedures to follow to achieve a reasonable representation of this type of marble. The base or ground colour should be black and a combination of greens are applied to the base, the veining is then added using feathers or brushes. The softener plays an important role in creating the correct effect.

Tools/paint	Method/technique
• lint free cloth • goose feathers • cotton buds • hog and badger softeners • fitches • paint palette • sponge Materials • prussian blue oil stain • yellow ochre oil stain • zinc white or eggshell white • raw linseed oil • white spirit • driers • transparent oil glaze.	• Apply gilp to the black base or ground coat (a mixture of two parts turpentine to one part linseed oil and 10% driers). • Load your palette an amount of each of the tinters, white eggshell and oil glaze. • Take a fitch and mix a dark green (brunswick) apply to the ground. • Repeat this process several times. • Mix a pale green and repeat the process. • The next stage is to open up the applied colour by spotting with the white spirit. • Soften with the softeners. • Now mix up the white and apply fine veins with the feather or sable brush. • Load the goose feather with this colour and draw it across the black ground twisting the feather through 180° from the wrist, this will create veins. • Use the softener at all stages. On completion of the effect apply a protective coating.

(a) (b)

Figure 10.20 *Vert de Mer marble examples (see Plate 12)*

Sienna marble

This marble is partially translucent and has a background ranging from cream through to yellow ochre. There are variegated patches of red, purple and green interlinked with veining that is random to the surface. The veins give a fissured look and fine veins can change from spidery to main arterial veins. Some areas are without veins and appear as pebbles in shape.

Tools

- lint free cloth
- cotton buds
- softeners
- fitches
- filberts
- lettering pencils
- palette board
- sponge.

Materials

- burnt sienna
- yellow ochre
- ultramarine
- indian red
- raw umber
- black
- white eggshell
- transparent oil glaze
- white spirit
- varnish.

Method/technique

- Prepare all tools materials brushes and equipment required for the task.
- The ground coat should be buff or cream.
- Mix up a gilp and apply to the surface.
- Mix up some glaze tinted with yellow ochre and apply to the surface with a fitch in wide angular veins. Link occasionally over the surface.
- Soften in all directions and leave to dry.
- Apply gilp over the whole surface and apply a glaze tinted with burnt umber, soften and wipe out areas with the cotton buds, eraser or cloth.
- Apply veining in varying widths, soften and allow to dry.
- Add more veins varying the colours, thicknesses and widths.
- Allow to dry and then varnish.

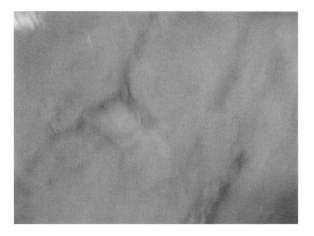

(a)

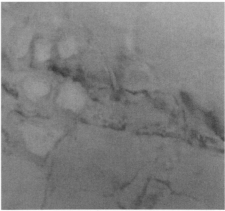

(b)

Figure 10.21 *Sienna marble examples (see Plate 13)*

10.5.7 Graining

Oak and mahogany
The graining skills required at this level of work are:

- brush graining
- dragging
- combing
- flogging.

To a prepared ground apply specially prepared media in the form of scumble. Do not over load the surface otherwise the effect will be difficult to create. Do not over thin with solvent or linseed oil otherwise the figure work created will flow out on drying.

Oak graining
Medium or dark oak scumble is a good media to practice with to create straight graining and combing in oak.

Stage 1 – apply the scumble to the surface and brush out evenly to produce a straight grain effect.

Stage 2 – obtain a drag brush and pull it through the produced straight grain whilst still wet. This will produce a more controlled straight grain effect.

Stage 3 – select a comb and through part of the dragged work apply the comb to the wet scumble and pull through the dragged effect, comb lines will be created producing a combed texture. Clean the comb, place on the surface of the combed work, slightly offset and comb again. This will create the pore effect.

Stage 4 – select a flogger and on a sample of work that has been dragged, hit the surface with the flogger from the bottom of the work to the top. A different wood texture will be created which represents timber pore structure.

Mahogany graining
Apply mahogany scumble to a prepared base coat or ground, brush out evenly and then flog from bottom to top removing all brush marks and replacing them with fine pore texture marks. This gives the base graining effect that can be left as a finished texture. Over this applied texture can be added the next stage of mahogany graining. Using the

Figure 10.22 *Figure work development for mahogany*

cutter apply a parabola shape to the wet scumble; keep increasing this shape until a section of heartwood has been achieved. Soften out from the centre to remove harsh brush marks. Finally use a mottler to place highlights.

10.5.8 Stencil work

The function of stencil work is to apply design and colour to selected surfaces to create decoration. The applied stencil can be a running border, a single motif or an overall applied pattern. Two or three dimensional effects can be created by clever use of stencil plate designs.

It is possible to convert any design motif into a stencil for application to a surface. Skills will be required in the drawing of the design, the conversion to a stencil and the cutting out of the plate itself. The final application of the colour through the plate to the surface will result in applied hand craft ornamentation.

There are two main types of stencil plate:

1.   positive where the design area is removed
2.   negative where the background to the design is removed.

Throughout the stencil should be connecting strips called 'ties'. Without these ties the stencil plate would fall apart. The tie should also look as if it is part of the design. Multi plate stencils are stencil plates made for various parts of the design, for instance, for each colour.

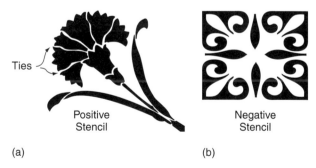

(a)   (b)

Figure 10.23 *Positive and negative stencil examples*

If the design requires leaves and flowers in two colours, one plate has the leaf design on it, the other the flower. Care must be taken to ensure that the two plates register with each other. For such work, careful marking out is required so that the lines on the surface match up with the lines or registration marks on the stencil plates. It is preferable to have two lines crossing at right angles on the stencil plate to indicate registration.

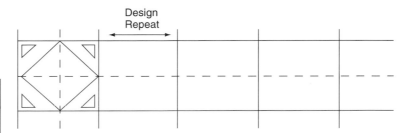

Figure 10.24 *Design repeat on a stencil run with registration lines*

Use of colours

The placing of negative stencils on surfaces such as friezes can create excellent effects. Stencil work texture is applied on the frieze and when the negative stencils are removed the background colour of the frieze shows through colours. Colour tipping is another useful approach. This is where the main colour is applied through the plate and other colours are applied at areas to create a multi colour effect.

Surfaces – when dealing with a dado and wall filling a band of colour may be painted to act as a dividing line. On this band a pattern may be stencilled on.

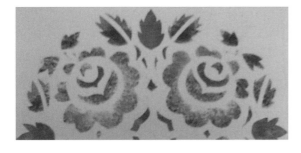

(a)

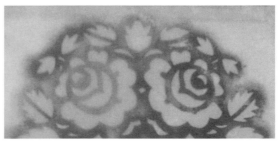

(b)

Figure 10.25 *Applied stencil positive and negative*

Application procedure

- Ensure that enough colour is pre-mixed to do the whole job.
- Have a stencil brush and palette available for each colour.
- Apply the selected colour to the palette.
- Tape up the bristles of the stencil brush to prevent them splaying during the stencilling process.
- Load the stencil brush with colour from the palette.
- Stipple the loaded brush on a clean area of the palette until colour distribution is even.
- Apply colour through the stencil plate design on to the wall surface.

- Allow each colour to dry.
- Apply alternate stencil applications to prevent plate contamination and smudging of the previous application.

10.5.9 *Gilding*

This is a specialist process and as such will be covered in depth in other publications. The information offered here is basic.

Gilding is the application of transfer or loose leaf gold or metallic foil to surfaces such as glass, plaster or timber ornamentation. The gold can be purchased in books of transfer or loose gold. There are usually 25 leaves to a book and prices are obtained on the day of purchase.

(a)   (b)   (c)

Figure 10.26 *Loose and transfer leaf foils*

Figure 10.27 *Gilders cushion, knife and tip (see Plate 14)*

(a)

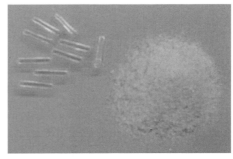
(b)

Figure 10.28 *Adhesives to apply gold*

Adhesives used to apply gold leaf
These are usually termed gilding sizes.

Japan gold size   This is used for gilding beadings, hollows or small work. It is available in various qualities:

- half hour
- one hour
- two hour
- four hour
- eight hour.

It is a quick drying varnish in simple terms, however it must be stated that gilding carried out on these sizes lacks lustre.

Silver leaf size   This requires a special size. The silver leaf quickly tarnishes and turns black if not protected. Silver leaf should be applied over quick Japan gold size that has little oil content. The leaf should then be lacquered.

Oil size   This is made from resins and drying oils and are adjusted to set in three, twelve and twenty four hours. They are also known as French sizes.

Gelatine size   This is made from boiling animal hoof, bone or skin. It is colourless and glutinous.

Isinglass   This is a pure form of gelatine made from the bladder of fish, an adhesive used for loose leaf glass gilding.

Key point

Gilding is a specialist subject and is covered in more detail in NVQ level 3 subject matter or in the Advanced Construction Award.

Test your knowledge 10.1

1. Name four categories of decorative effect (not individual treatments).
2. What base coat or ground coat is most suitable for the application of oil based decorative treatments?
3. Describe the purpose of gilp and how it is mixed.
4. State the two media that are used to apply decorative treatments.
5. Name two treatments that can be applied to a surface without the requirement of a brush during the texturing stage.
6. What is the name given to the part that holds the stencil plate together?
7. State the effects that can be produced during graining processes.
8. Name the materials that can be used to produce stencil plates.
9. List the brushes, tools and materials required to produce a piece of marble.
10. Describe the process of rag rolling.

Place your responses in your portfolio as evidence.

11 Scaffold

11.1 Introduction

The scaffold in this chapter will be access scaffold only, up to a platform height of 2 m. No person or persons should use access scaffold unless previous training has been given and that person has been assessed as competent in the use of such access scaffold. Technical detail will be offered but this can change as legislation changes.

The information offered in this section is for guidance only.

Access scaffold include:

- ladders both timber and aluminium
- stepladders both timber and aluminium
- trestles
- batons, boards and staging.

11.2 Ladders

Timber ladders include:

- Standing ladders are single section ladders up to 7.3 m long.
- Double extension ladders are two sections similar to the single standing ladder, but are connected by brackets and guides. Without ropes the ladders are available up to 4.9 m in length when closed. Rope operated types are available in closed lengths of up to 7.3 m.
- Pole ladders are single section ladders where the stiles are made from one tree cut down the middle. This ensures even strength and stability. Obtainable in lengths up to 12 m and mainly used as access ladders to tubular scaffolding.

Scaffold parts

- *Stiles* – usually made from Douglas fir, redwood or hemlock. The stiles house the rungs.
- *Rungs* – round or rectangular in shape constructed from oak, ash or hickory.

- *Ties* – steel rods fitted below the second ring from each end of the ladder and not less than nine rung intervals. Good quality ladders can have ties fitted under every rung.
- *Reinforcing wires* – steel or galvanised cable is fitted into grooves of the stiles to give extra strength.
- *Guide bracket* – fitted to the top of the lower section of extension ladders. The top section of the ladder is slid and located under the guide.
- *Latching hooks* – fitted to the bottom of the top section to hook over the section below when raised to required height.
- *Pulley wheels* – act as guides for the movement of the ropes when raising and lowering extension sections.
- *Ropes* – hemp sash or cord or other material of equivalent strength which offer good grip when raising or lowering ladder sections.

Aluminium ladders – are lighter than timber ladders due to the alloy used to construct them. They are extremely strong, will not rot twist or warp. The rungs must be non-slip and be spaced at 230 to 250 mm intervals. The ends of the stiles must have non-slip pads fitted.

Aluminium ladders include:

- standing ladders up to 9 m in length
- extension ladders up to 12.5 m in length
- system designed ladders that can be erected to act as access to staircases or to provide a platform on varying levels.

Scaffold parts

- stiles are in box section or shaped to slot together
- rungs are pressed to the stiles in box section with anti-slip grips
- location clips or guides are wrap round
- non-slip pads are fitted to the base of the stiles
- automatic pawls are fitted to the more expensive ladders in place of the latching hooks. These pawls automatically engage as the ladder is raised.

11.2.1 Raising and lowering ladders

Extension ladders should be raised with the sections closed. If long extension ladders are to be used, raise one section at a time, slot them together and then extend to the correct working height. Two persons will be required to raise and lower the larger ladders. The following is the procedure:

- Lay the ladder flat.
- One person stands and foots the bottom rung and steadies the stiles as the ladder is to be raised.

- The second person stands at the top end of the ladder and begins to raise the ladder over their head walking towards the base of the ladder. The rungs are used one by one to raise the ladder vertically.
- The ladder is then rested against the wall surface and extended.
- Finally the collect ladder angle is achieved to a ratio 4 up to 1 out.
- Or a safety angle of 75°.
- Reverse the process when lowering the ladder.

11.2.2 Lifting and carrying ladders

To transport ladders manually over short distances, usually along the elevation of the property you are working on, the procedure to follow is:

- Lift the ladder away from the wall to its vertical position.
- Grasp a rung and stile within easy reach, do not over stretch.
- Raise the ladder from the ground and find the balance and angle of balance.
- Move the ladder by walking a few paces to the next required position on the wall elevation.
- Under no circumstances try to move a ladder that is fully extended in this manner.
- If the ladder has to be moved more than a few metres, lower the ladder and two persons should carry the ladder to the required site, either on the shoulders or by the side.
- Both persons should be on the same side of the ladder otherwise when a turn is made the pressure of the ladder becomes uncomfortable for one carrier.

Safe working procedure
Ladders must project 5 rungs (1 m) above any working platform when erected for access.
Never:

- stand ladders on uneven or loose ground unless secured correctly
- stand ladders on unsteady bases
- use makeshift ladders
- use a ladder which will not reach
- wear inappropriate footwear
- carry too much equipment up a ladder
- over reach when working from ladders.

Take extra care when raising ladders near overhead electric cables. For overnight safety ladders should normally be lowered at the end of each working day and stored securely. If this is not possible a 2 m long scaffold board should be lashed to the ladder to prevent access to the rungs. Warning signs should also be posted, 'Do not use this scaffold'.

11.2.3 Tying ladders

The construction working places regulations state that ladders must have a firm and level base on which to stand and if over 3 metres long they must be fixed:

- At the top or the bottom
- If neither is possible a person must foot the ladder. That is to stand with one foot on the bottom rung and the other firmly on the ground. Both stiles must be held and attention be paid at all times.

Lashings must be tied to secure positions such as scaffold tubes or screw eyes which have been firmly screwed into walls. Never tie on to gutters or pipes. Never create a tripping hazard on rungs or platforms when tying scaffold.

11.3
Stepladders

Only stepladders constructed from timber and aluminium are covered in this chapter. There is glass reinforced polyester types available on the market but are not generally used in the trade. Stepladders can be purchased as trade, industrial or domestic quality, each has its limitations as to use and the loading that can be placed on them. Labels are fixed to stepladders to indicate usage requirements and weight limitations.

Stepladder parts

- *Stiles* – taper from bottom to top and are wide enough to accommodate one 250 mm scaffold board or baton.
- *Treads* – are at least 90 mm deep and spaced at 250 mm intervals vertically.
- *Tie rods or reinforcing bars* – stepladders can be fitted with either to strengthen the whole structure. The reinforcement can be fitted under every tread.
- *Back support* – hinged and bolted to the back of the hanging board. This is a stabilising support and should never be used to carry weight.
- *Locking bars* – fitted to aluminium steps. These must be fully extended and placed into the locked position when using stepladders.

- *Feet* – must be fitted to aluminium stepladders to prevent movement and damage to floors.
- *Ropes or cords* – should not be less than 6.5 mm in diameter, two lengths should be fitted to each stepladder and the cord must be knotted at stiles and stability support. The stepladder must open to the correct angle for working (65–75°).

Stepladder sizes
Stepladders are sold by the number of treads required. They can be purchased with up to 14 treads. The examples listed are sizes up to a 2 m platform. The top counts as one tread.

- 6 tread – 1.3/1.4 m
- 8 tread – 1.7/1.8 m
- 10 tread – 2.1/2.3 m

Never use the top four treads of a stepladder unless a handrail is built in.

Figure 11.1 *Latching hook*

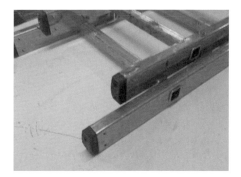

Figure 11.2 *Double extension ladder detail*

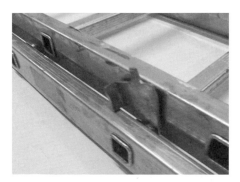

Figure 11.3 *Latch release clip*

Figure 11.4 *Foot detail*

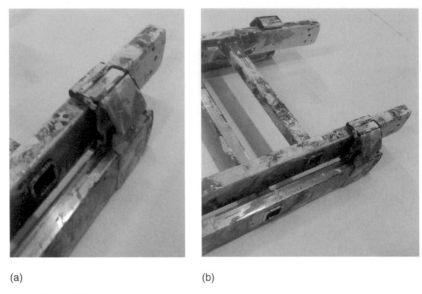

(a)   (b)

Figure 11.5 *Location guide detail at ladder top*

11.4
Trestles
(folding)

Folding trestles are constructed from softwood timber or metal such as aluminium alloy or steel.

- They are tapered from top to bottom and the stiles are wide enough apart to take two 250 mm scaffold boards or a lightweight staging.
- Cross bearers (similar to rungs or treads) are staggered on either side of the trestle to give platform rises of 230 mm–260 mm.
- Tie bars are filled to each leg of the trestle in at least two places. This gives the trestle additional stability.
- The hinges used on the trestle are lipped and when fully opened lock into position.
- Sizes available vary according to manufacturers, check catalogues.
- Regulations state that toe boards and handrails must be fitted to platforms 2 m or above.

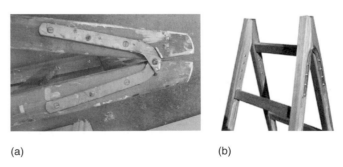

(a)   (b)

Figure 11.6 *Locking hinge detail*

Erecting, lifting and carrying trestle scaffolds

- When raising trestles lay them on edge and slightly open the legs. Lift the hinge or top of the trestle until vertical. Repeat with the second trestle, position where required and add the platform. Make sure that the trestles are both facing and the cross bearers matching.
- The platform must never have a trap end created. The overhang of the platform should be four times the thickness, no more, no less.
- When moving trestle scaffold dismantle and re-erect.
- Never lift and drag erected trestle scaffolds.
- To prevent damage when moving trestles ensure the check blocks are engaged. These blocks prevent distortion or pressure being placed on the hinge.

Hop ups – are purpose made timber scaffolds that allow a working height of approximately 2.4 m to be reached. The use of two hop ups with scaffold boards provides a suitable access scaffold for low work.

- The top platform should be 500 mm × 500 mm.
- The height is approximately 500 mm with one step each end of 250 mm wide and high.
- The base dimension of a hop up is 1 m × 500 mm.
- Before use check hop ups to ensure stability.

11.5 Boards and staging

Scaffold boards should be constructed to comply with British standard Specification. The ends of scaffold boards are either cut at a taper or galvanised metal edge bands are fitted. Into this band is pressed the kite mark and BS conformity number. The timber used for the construction of the boards should be straight grained and free from knots, splits or shakes.

Figure 11.7 *Protective edge to board*

Key point

Board overhang – no board should overhang its supports by more than four times its thickness, or less than 50 mm.

Board dimensions are not usually longer than 4 m unless they are specials, the thickness of the board relates to its length

- 40 mm thick up to 2.750 m in length
- 50 mm thick up to 3.000 m in length
- 75 mm thick up to 4.000 m in length
- boards less than 50 mm thick must be 200 mm in width
- boards more than 50 mm thick can be from 150 mm wide.

The distance that boards can be spanned before support is required are as follows:

- 40 mm thick boards require a support at every 1.5 m
- 50 mm thick boards require a support at every 2.5 m
- do not use two scaffold boards one on top of the other.

Lightweight staging These provide a wider working platform than batons and can be used for spanning greater lengths without support. They are usually used with trestles or towers. The staging is designed to take a loading of three persons together with lightweight equipment of up to 272 kg, if evenly distributed. Sizes are available from widths of 450 mm or 600 mm with lengths from 3 m to 7.2 m. Full handrail systems are available for this product.

Timber staging construction

- The stiles are reinforced with galvanised high tensile steel wire.
- The ends of the stiles are protected with hooping irons.
- Cross supports are fitted every 380 mm or 450 mm centres along the length of the staging.
- These cross supports are reinforced with steel rods.
- Walking slats are double screwed at each cross rail.
- Safety tie hooks are fitted to allow lashing of staging when in place to prevent movement of the staging.
- Store flat on evenly spaced floor batons when not in use.

 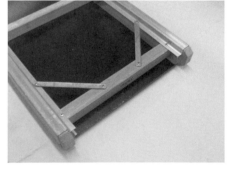

(a)   (b)

Figure 11.8 *Lightweight staging showing detail on construction*

11.6 Safety checklists

Key point

Safety checklist for ladders

Timber ladders
Do not use if:

- there are broken, missing or damaged rungs
- there are broken, or repaired stiles
- there are defective ropes, guide brackets, latching hooks or pulleys.

Ladders must be:

- straight grained and free from knots or resin pockets
- free from splits, shakes and decay
- rungs should be clean with no dirt or grease present
- free from paint, never paint timber scaffold.

BS 1129/66 is the specification number for timber ladders

Aluminium ladders
Do not use if:

- there are broken or distorted rungs
- there is any visible damage to the structure of the ladder or its parts
- rungs should be clean and free from dirt or grease.

Warning!

- aluminium ladders should not be used near to low, medium or high voltage power cables
- aluminium can be damaged by exposure to chemicals present in corrosive atmospheres.

BS 2037 is the specification number for aluminium ladders.

Key point

Safety checklist for boards and staging

All timber products must:

- be constructed from straight grained timber
- be free of knots, shakes, splits or decay
- be clean and free from dirt and grease
- never be painted
- boards should not be used if twisted or warped or have split ends
- staging must have reinforcing wires and support bars present.

BS number for Scaffold boards is BS 2482/70. BS number for lightweight staging is BS 1129/66.

Test your knowledge 11.1

1. What is the correct angle or ratio used when erecting ladders?
2. How far above a working platform should a ladder extend?
3. List some safety precautions to take when using ladders.
4. What checks should be made before using stepladders?
5. Where are tie rods fixed on stepladders?
6. What is the distance between rungs on extension ladders?
7. What should never be applied to timber scaffold as protection and why?
8. When using extension ladders what should be the minimum rung overlap for a ladder extended up to 4.8 m?
9. What device prevents a trestle from collapsing?
10. How can scaffold board ends be protected from possible damage?
11. On 50-mm thick scaffold boards how often along their span must they be supported?
12. What is the maximum and minimum overhang of a scaffold board?

Place your responses in your portfolio for future evidence.

Key point

Safety checklist for stepladders and trestles

Timber stepladders
Do not use if:

- there are broken, missing or damaged rungs
- there are broken, or repaired stiles
- there are defective ropes, guide brackets, latching hooks or pulleys
- back flap hinges are loose
- screw or bolts are loose.

Stepladders must be:

- straight grained and free from knots or resin pockets
- free from splits, shakes and decay
- rungs should be clean with no dirt or grease present
- free from paint, never paint timber scaffold.

Aluminium stepladders
Do not use if:

- there are broken or distorted rungs
- there is any visible damage to the structure of the ladder or its parts
- treads are dirty or contaminated
- the locking bars are damaged
- feet stops are missing.

Folding trestles
Do not use if:

- cross bearers are damaged or loose
- stiles are damaged or makeshift repairs have been carried out
- hinges are worn, loose and do not lock correctly
- splits or decay are present.

12 Health and safety

12.1 Introduction

Safety is every person's concern. There are not often second chances if a person is injured through some other person's negligence, fooling around or general ignorance. To offer guidance, regulations were introduced named 'The Health and Safety at Work Act 1974' (HASWA), all legislation in the act is enforced by law.

The organisation that is responsible for the enforcement is the 'Health and Safety Executive' (HSE). The people who work for the HSE are commonly known as Factory Inspectors or Health and Safety Inspectors.

The powers and duties of the HSE allow the:

- entering of premises at any reasonable time (day or night) if a dangerous situation is occurring
- examination and investigation of situations as deemed necessary
- collection of evidence either manually or electronically
- examination of relevant documents
- making safe of dangerous equipment
- questioning of persons and declarations of truth to be signed
- use of the police force to effect entry to premises.

Notices of enforcement If correct safety law is not followed and rules have been adapted to allow working procedures to take place, a dangerous situation could occur and could possibly lead to accidents. The HSE uses notices to prevent this practice and if it suspects there are such situations it can issue a:

- *Improvement notice* – if an inspector considers safety law is in contravention of legislation, an improvement notice is issued to the company and that person responsible for safety in the company is given a time span to act upon such contraventions.
- *Prohibition notice* – if it is considered that dangerous working practices are taking place, an immediate prohibition notice can

be issued which stops such working practices until remedies have been implemented.

- *Appeals* – if a company disagrees with an issued notice, an appeal must be received by the HSE in writing within a period of 21 days of the notice being served.

HASWA duties   The setting of standards, a quote from the act states 'it shall be the duty of every employer to ensure, so far as is reasonably practicable the health, safety and welfare of all employees'.

HASWA requires duties to be carried out by individuals. Failure to meet these requirements is a criminal offence in law, and fines or imprisonment can result. They may be:

- employees duties
- employers duties
- individuals duties.

12.2 Duties and responsibilities

12.2.1 Employer's duties

- Provide and maintain plant, machinery, equipment and ensure safe systems of work are adhered to.
- Provide training, give instruction, produce information and implement supervision in the workplace to maintain the health and safety of employees.
- Provide and maintain a safe place of work without risks to health and welfare.
- Set up procedures for handling, storage and transport of articles and substances in compliance with HASWA.

Plant and equipment

- Is it checked regularly?
- Who carries out repairs?
- Are records of maintenance and repair kept?

Employee's awareness

- Do they know HASWA legislation?
- Have they received training to use equipment both old and new?
- Can they use safe working practices?
- Do they comply with personal hygiene requirements?
- Are they aware of hazards and dangers?

Lifting carrying handling and storage

- How many employees have back complaints?
- How is scaffold, plant and machinery transported?
- Are all employees aware of correct lifting carrying and handling techniques and has training been given?

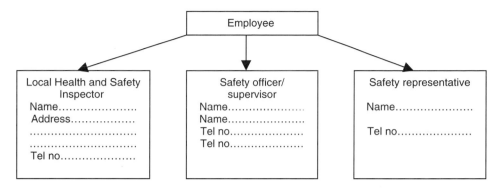

Figure 12.1 *Contacts relating to safety at work*

12.2.2 Employee's duties

- To take reasonable care of yourself and others that could be affected by what you do related to HASWA.
- To cooperate with employer or employer's representative in all matters relating to HASWA.
- Do not misuse or interfere with anything provided for your health and safety.

Awareness

- Am I aware of the company safety policy?
- Do I know of the risks related to my employment?
- Can I work safely without causing danger to others?
- Do I wear the correct PPE?
- Do I follow safety codes of practice at work?
- Have I received the training for the tasks I carry out at work?
- Are my tools and equipment safe to use?
- Do I know who is responsible for me in my workplace?
- Do I know the name and contact number of my safety officer or supervisor?

If the answer to any of the above is 'no', then find out by asking your immediate supervisor, instructor, lecturer or employer. The law requires you to know your duties under the HASWA.

Safety Policy   If an employer has more than five employees, a statement of safety policy is required and a copy should be given to all employees. It is also required by other companies if you sub contract work.

The policy should include:

- a statement of intent, that is what the employer intends should happen in the company
- comprehensive information should be available as to how the company is to comply with health and safety law in the setting up of its working environment

- what health and safety responsibilities exist
- safe systems of work, how and where they are described
- who is responsible for delegated duties
- keeping of records
- arrangements for reviews, updates and training.

Safe systems of work   A code of practice should be available for activities carried out during the execution of work for the company. The code should relate to:

- people and their behaviour
- equipment, its use and condition
- the environment in which the employees work.

When formulating a code of practice take into account risk assessment.

12.2.3 Information and training

Communication is required to ensure that a contract progresses satisfactorily and safely from the employer through the managers to the supervisors to the workforce.

HASWA requires that training, supervision and information is readily available for all employees to enable safe work practices to be carried out.

Here is a training breakdown:

- Essential information – safe working systems, safety policy statements and safety posters
- Job safety – to include on the job training and off the job training
- Awareness and training related to HASWA
- Supervision – to include effective training in safe working practices and procedures.

12.3 Inspections and investigations

Safety inspections   It is the responsibility of management to ensure that inspections are carried out at regular intervals in the workplace. Management representatives carry out these in house checks for the company or on behalf of the HSE.

The inspection will look for:

- poor housekeeping
- causes for concern
- poor working practices
- possible hazards in the workplace
- dangerous occurrences.

The inspection will also check the following:

- machinery guards are in place and comply with legislation
- equipment

- condition of scaffold and scaffold inspection reports
- stage one safety inspections, hazard spotting exercises using checklists, to be carried out monthly
- risk assessments carried out and up to date
- first aid book and records kept.

On completion of a safety inspection an action plan should be drawn up and actions carried out to compliance within a given and reasonable timescale.

Accident investigation   The following persons will investigate accidents that may occur in the workplace:

- employer's representative
- safety officer
- senior manager
- solicitor
- local health and safety officer
- trade union representative
- insurance company.

It will involve visits to the site of the accident, the collecting of evidence in the form of statements, witness testimony, photographs. Negotiations will then take place between all parties to resolve the situation.

An accident is an unplanned mishap; it can result in damage to persons or property. The implications to the employer:

- the loss of labour
- insurance premiums could rise
- faces prosecution or civil action
- worry and stress
- loss of business
- gains a bad reputation.

The implications to the victim:

- loss of life
- enduring pain and suffering
- can suffer a disability
- loss of livelihood
- financial loss
- prosecution.

Prevention of accidents can be reduced if you are aware:

- understand what could and can go wrong in the workplace
- work safely
- have a good attitude
- only use tools and equipment that are safe to use
- use correct working techniques.

Key point

Stage one inspections should occur regularly, at least once a month to check the safety of the workplace. The inspection looks at the working environment and allows reports to be actioned regarding maintenance of the workplace. It could be anything from blown light bulbs to empty fire extinguishers.

12.4
Welfare

Protective clothing should be issued by the employer to the employee and this issued clothing must be worn at all times.

- head protection – safety helmet or bump hat
- eye protection – safety goggles, glasses or visors
- foot protection – safety boots, shoes or trainers
- body protection – overalls, gloves, gauntlets, aprons, ear defenders, respirators or dust masks.

It is the responsibility of the employee to present themselves in a manner capable of producing work. Alcohol or drug related substances could affect your ability to work safely; they can also contribute to accidents, 'You must be fit for work'.

Facilities, depending upon numbers working for a company the following could be provided:

- toilets
- washing and changing facilities
- lockers
- places to eat
- supply of drinking water
- first aid facilities.

Notices should be placed in locations such as notice boards or where personnel assemble to make the workforce aware of current information.

- safety signs
- evacuation procedures
- exit direction signs
- copies of regulations
- assessment sheets
- contact names and numbers.

There should be equipment to use if required in the working area such as:

- fire extinguishers
- first aid boxes
- special equipment.

      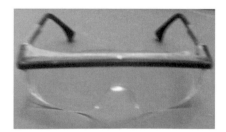

(a)   (b)   (c)

Figure 12.2 *Items of PPE*

12.4.1 Hazardous materials and their influences

Building sites and working environments in the construction industry can present the worker with many hazards.

- *Dust* – can enter the body and cause cancer, irritation, dermatitis, ulcers, stomach problems and bronchitis. Fumes can be inhaled. It may not appear to be a problem when young, but the disease can develop over a period of time, 'Wear your respiratory equipment'.
- *Asbestos* – This is still present in many old buildings as it was used extensively for insulation against fire, noise and heat. If asbestos is uncovered whilst working on site, work must stop and special teams are brought in to deal with the removal. No person can work with asbestos unless they hold a special licence issued by the HSE. Asbestos dust is a killer.
- *Lead* – This is a material that you would never come into contact with, or is it? Consider lead flashings, the sanding and abrading of old paintwork, fumes created from burning off old coatings that may contain lead products. Refer to the code of practice issued by the HSE that covers the control of lead in the workplace.
- *Radiation* – Seek advice on safety procedures before entering premises bearing the yellow and black warning triangles. Special protective equipment may have to be worn. These are usually restricted working environments.
- *Noise* – The noise at work regulations 1989 sets noise levels and exposure levels, it places duties of responsibility with employers and employees. Communication can be impaired which puts workers at risk therefore safeguards must be implemented to account for noisy working environments.
- *Vibration* – The use of tools and equipment can cause injury to employees through continuous use of such equipment. Even vibration through buildings can affect workers. The blood supply can be affected which results in an industrial disease called 'white finger'.

Test your knowledge 12.1

1. Name the organisation responsible for enforcing safety law.
2. What powers do they have?
3. What are the responsibilities of the employer to the business and the employees regarding health and safety?
4. What are the consequences of an accident to the employer and employee?
5. Describe 'HASWA'.

File your responses to use as evidence.

12.5 COSHH

The control of substances hazardous to health regulations 1988 known in short as COSHH are a series of regulations formulated by the Health and Safety Executive. The regulations came fully into force on 1 October 1989. The purpose of the regulations is to protect people in the workplace from the effects of substances hazardous to health. Failure to comply with COSHH regulations is an offence and is subject to penalties under the Health and Safety at Work Act 1974.

12.5.1 COSHH hazards in painting and decorating

In Painting and Decorating there are many hazards due to the various materials used. It is often ignorance of the material and of how to use it that can cause health problems either immediately or in later years.

Substances hazardous to health include those with labels on their packaging stating a hazard exists such as:

- highly toxic
- skin irritant
- corrosive substance
- highly inflammable.

This is a warning and it informs the user operative of a hazard to be aware of.

Other hazards are not so obvious.

- we touch liquids (spillage, cleaning of hands, face, etc.)
- we inhale dust and fumes.
- we absorb chemicals into our skin by touch.

All of the above constitute a hazard to:

- persons in direct contact with the material
- persons in the vicinity unaware of the presence of the material.

Painter's activities – every day activities that a painter and decorator might carry out:

- preparation of surfaces
- preparation of materials
- application of materials.

These tasks do not look too hazardous do they?

- Sandpapering, plaster, filler etc., *hazard* – dust inhalation.
- Using paint stripper, *hazard* – fumes, highly flammable, skin irritant.
- Preparing paints, *hazard* – fumes from material and solvents.
- Mixing adhesives, *hazard* – dust inhalation, skin irritant.
- Application of paint by spray, *hazard* – fumes, dust (spray) flammable.

Respiratory Problems – from dust and fumes can cause:

- bronchitis
- asthma
- asbestosis.

By touch during preparation:

- lead poisoning
- colic
- dermatitis.

Ailments
If you want these ailments, carry on with poor working practices. They won't cost you anything except possibly your **health** and finally your **life**.

For *good health*

- look after yourself
- follow recommendations
- use good working practices
- observe COSHH.

12.5.2 COSHH hazard and risk

A hazard
The hazard presented by a substance or material is its potential to cause you harm. It may make you:

- cough
- damage your liver or
- kill you, if you breathe it in, swallow it or get it on your skin.

A risk
A risk from a substance is the fact that it will harm you when using, it will depend on:

- the hazard presented
- how it is used
- how it is controlled in use
- who is exposed
- to how much
- for how long.

Control
Poor control of substances can create risk, even from low hazard substances and materials. But even the most dangerous substances can be used safely with the correct equipment and technique:

- adequate ventilation
- correct safety wear
- correct safety equipment.

12.5.3 Employers and employees responsibilities

Taking into account the type of work undertaken by a firm, the employer must determine the hazards of substances used to carry out such work for the firm.

Employers duties
- assess the risks to people's health from the way the substance is used in the workplace
- prevent anyone from being exposed to the substance
- decide how to control exposure to reduce the risk
- establish effective controls
- train and inform the work force.

Additional duties
The employer needs to monitor employee's health and exposure to such substances.

12.5.4 Employee's duties in identifying hazards

The employee needs to know more than the fact that a hazard exists. The employee should be trained to understand these hazards and risks when using such substances. The employee has to know:

- What the risks are from using such substances in the workplace.
- How the risks are controlled.
- What precautions have to be taken when using such substances.

Other legislation
COSHH does not apply to any work done with:

- lead
- asbestos
- radioactive materials.

They are covered by other legislation regulations.
Substances that are hazardous include:
At work:

- cements
- plasters
- paints and varnishes
- thinners and solvents
- oils and grease.

Others:

- adhesives
- glass fibres
- weed killers
- abrasive dusts.

In situ hazards
Substances can also be on site before you arrive (toxic or poisonous).

- in the soil
- bird droppings

- silicas in masonry
- substances left by others
- old quarries, tips, etc.

Six simple steps to remember

1. Have knowledge of the products and substances you are using.
2. Assess the hazards to health that the substances can cause.
3. Eliminate or control the hazard.
4. Obtain information. Give instruction and training to employees and people who could come into contact with such substances.
5. Monitor and control people using such substances.
6. Keep records of where such substances were used, by whom and when. Some records may have to be kept for up to 30 years.

Test your knowledge 12.2

1. Which safety body introduced the COSHH regulations?
2. What was the purpose of the regulations?
3. State three regulations, relating to COSHH that are the employer's and employee's responsibility?
4. Identify an activity carried out by a painter and decorator that would constitute a health hazard.
5. Name three materials that are not controlled by COSHH, but have their own regulations.
6. Identify three items of equipment that could be worn to prevent contamination when working.
7. Identify records that should be kept by employers for reference.
8. Who is responsible for the implementation of safe working practice in the workplace?

13 Glossary

Agitator Device for mixing paint that has settled in a container.

Adhesion Ability of coating to stay on the surface by specific or mechanical key.

Alkyd Type of synthetic resin derived from acids or alcohol.

Batch number A number given to a run of wallpaper at manufacture.

Bituminous paint Coatings based on natural bitumen and coal tars. They were used frequently on metal rain ware such as guttering and down pipes.

Bleeding The ability of a substance to come through a coating applied over itself.

Blistering The formation of pockets of resin or liquid as they leave surfaces and push off the applied coating.

Blooming A film that forms on the surface of coatings when there is sudden temperature change. This bloom can sometimes be removed by wiping with a damp cloth.

Bring forwards A term used to describe the initial preparation and application of coatings up to a readiness for the application of the final or finishing coating.

Broken colour A term used to describe the application of colour to a surface which is then moved about to produce various textures.

Build The paint system itself representing film thickness of an applied system.

Cellulose paste An adhesive based on ethers in dried form. When water is added a wallpaper paste is produced.

Chalking A defect where the vehicle or resin in the paint film has decayed due to weather influences. The pigment in the paint film now appears as a powdery deposit and can be wiped from a surface. Some paint coatings are applied with this function to act as a sacrificial coating during weathering.

Cissing A defect that occurs when an applied coating does not bond with a surface. The paint recedes and leaves areas of the surface bare, in some cases appearing as pin holing.

Clearcole Mixture of glue size and white lead, traditionally used as a primer for distemper (water based paint).

Coat   A term that describes the application of one layer of a system.

Consistency   The thickness or thinness of a liquid when mixed.

Covering power   The ability of a coat of paint to cover or obliterate the coating underneath. A loosely used term.

Cracking   The breaking up of an applied coating caused by the coating being too brittle, becoming denatured due to excessive surface movement. Crazing and checking are similar defects.

Contaminant   Dirt, dust or bits in materials.

Decant   To open and pour into a container.

Dispersed   Spread out evenly.

Distemper   Non-washable paint in which glue size is usually the base. Making a comeback as fashion paint.

Drop pattern   A wall covering whose match does not meet in a horizontal line equally on both edges.

Dutch metal   A copper based aluminium leaf used as a cheaper alternative to gold leaf.

Efflorescence   A white crystalline deposit left on the surface of plaster, bricks and concrete after the evaporation of free moisture. The salts are alkaline in nature.

Eggshell   Paint with a semi sheen or shine.

Embossed paper   Wallpapers that have relief texture after being passed through embossing rollers during manufacture. Care should be taken not to oversoak or over brush during pasting and hanging as the emboss will be lost.

Extenders   Pigments that become semi transparent when ground in oil. They are a cheaper form of pigment used to bulk out paints to make them more affordable.

Fat edges   The build up of paint created when joining wet paint when the application speed has been too slow.

Fillers   Materials in powder or ready mixed form used to repair minor cracks and indentations in surfaces prior to the application of coatings.

Flaking   The lifting of the paint film in small areas representing small scales, leaving the under surface exposed to weathering.

Flashing   The effect of different sheen edges created when applying paint coatings, usually eggshells.

French polish   Mixture of shellac and methylated spirits.

Fungicide   A substance added to products that destroys the ability of moulds and mildew to flourish.

Gauze   A fine mesh trough in which liquids are strained.

Glazing   A decorative coating to which tinters or stainers are added. It is a semi transparent coating that can be applied to a surface and then be textured.

Gloss   Finishing paint with high reflectance value.

Gold leaf   Pure gold beaten into thin leaves in readiness for application to surfaces.

Gold size   This is the adhesive used to affix gold leaf to a surface.

Graining   A decorative treatment produced using special paints and brushes to imitate real timber.

High build   Viscous paint, producing thick coatings.

Kettles   A paint container.

Knotting   A spirit solution used as a sealer of resin seepage.

Laying off   The spreading out of the paint film in the final direction.

Lifting   The effect the next coat of paint has on the previously applied coat. The solvent in the paint acts like a paint remover.

Lining paper   A preparation paper used under wallpaper to create a clean surface of even porosity.

Linseed oil   A drying oil obtained from the flax plant.

Long oil   Denotes that the amount of oil in a varnish would not be less than 70%.

Making good   Preparing defective surfaces for redecoration.

Medium   Type of vehicle (liquid) in which the pigment is dispersed.

Opacity   Ability of paint to hide a surface or previous coating.

Orange peel   A paint finish created by using rollers or spray guns. The finished effect is considered a paint defect.

Paint remover   A liquid chemical that softens dried paint films to enable removal for redecoration.

Pigment   Material which provides colour and covering power within paint.

Polymer   Synthetic resin in which large molecules are built up.

Pot life   The time a paint coating remains usable after mixing two-pack coatings together.

Primer   First coat applied to bare or new surfaces.

Putty   A stopper manufactured from whiting and linseed oil.

PVA   Polyvinyl acetate, commonly used as a binder in emulsion paint.

Recoat able   The time when the next coat of paint can be applied without the solvent in the coating re-activating the first coating.

Relief decoration   Wall coverings that have raised patterns or textures.

Resin   Originally naturally occurring gums, soluble in natural solvents and oils, now largely replaced by man-made materials, especially alkyds with additional properties.

Rollers   Cylindrical devices that are used to apply coatings to surfaces.

Ropiness   Brush marks that are visible after applying thick coatings that do not flow out after application.

Round   A term used to describe a thick heavily bodied paint with good obliterating properties.

Runs   A defect where applied paint flows after application due to overloading.

Sealer   Usually a clear coating applied to an absorbent surface or powdery surface to seal it.

Sanding   A loose term used to describe the abrading of a surface.

Saponification   The effect of free alkaline softening and discolouring applied oil based paint coatings.

Scumble   A heavy pigmented medium used for graining.

Set pattern   The matching of wallpaper where each edge of a wallpaper marries up without adjustment of the wallpaper.

Sediment   Solid content in a solution.

Settlement   The settling of the pigment in a paint solution to the bottom of the container.

Sharp coat A thinned paint applied to even out the porosity of absorbent surfaces.

Shade number A code on wallpaper labels indicating the colours used during manufacture.

Size A solution of glue or wallpaper adhesive used at the correct consistency and applied to porous walls to allow slip and slide of wallpaper during application.

Skin The dried surface of liquid paint in a container.

Soaking time The time that is allowed to pass after applying wallpaper adhesive to wall papers prior to applying the wallpaper to the surface.

Solvent Part that evaporates as the paint is drying.

Spreading rate The area paint will cover per litre.

Staining The colouring of paint with tinters in the container.

Stippling The texturing of applied coatings with special brush to remove brush marks.

Sundries Articles that aid with the decorating process.

Tac rag A muslin or bandage-like material impregnated with drying oil. It is used to remove dust from surfaces before application of coatings.

Thinners A liquid that disperses pigment in paint.

Touch dry Term used to convey that the surface of the paint is dry but the under film has not hardened.

Undercoat The build up coat of a paint system, it covers and obliterates the surface prior to the finishing coat.

Varnish A clear coating applied as a protective cover to surfaces.

Viscosity The consistency of coatings and their ability to flow or resist flow.

Vehicle Liquid part of the paint.

Wet edge The area where one application of paint meets the next applied area and will merge without leaving a visible line.

Wood stain A decorative liquid that alters and enhances the appearance of timbers.

Yellowing The effect that light has on applied coatings over a period of time.

Chapter 1

Test your knowledge 1.1

Q1 The hand tool used to remove paint from moldings is a combination shave hook.

Q2 Sweeping brush and shovel or a hand brush and dustpan would be used to clean up general debris from the work station.

Q3 A scraper or broad knife would be required.

Figure 14.1

Figure 14.2

Q4 A filling knife would be used to apply proprietary filler to a surface during the preparatory stages. A stopping knife could be used to apply stopper or putties to window rebates.

Figure 14.3

Figure 14.4

Test your knowledge 1.2

Q1   The selections required are the plumb bob and line plus the spirit level, illustrations 1 and 5.

Q2   Illustration 6 is a wallpaper scourer used to scratch the face of hung wallpaper to aid penetration of water prior to soaking and removal

Q3   Illustration 5 is a spirit level used to determine pure horizontal and verticals during marking out of or hanging wallpapers.

Q4   Rules and measures should never be used as straightedges (Figure 1.28).

Test your knowledge 1.3

Q1   shave hook – (d)

Q2   spirit level – (c)

Q3   transformer – (c)

Chapter 2

Test your knowledge 2.1

Q1   Previously painted surfaces in good condition should be washed down to remove dirt and grease using a solution of sugar soap or mild detergent and warm water. Rinse with clean water to remove residue and then abrade using the wet and dry process or the dry process as required.

Q2   Equipment that can be used to aid the preparatory process are sponges, buckets, dust brush, brushes and shovels, scrapers, steam strippers, filling knives, shave hooks, electric or pneumatic sanders.

Q3   The breakdown of paint systems can be due to the following:

- wear and tear of everyday environmental traffic both internal and external
- poor initial preparation
- using cheap products
- vandalism
- industrial, coastal or rural atmospheres

Q4   Climatic conditions contribute to the breakdown of paint systems mainly due to extreme temperature changes and chemicals in the atmosphere such as acids and alkalis get deposited on surfaces.

Test your knowledge 2.2

Q1   The classification of timber is 'hardwoods' or 'softwoods'

Q2   Softwoods due to their resinous nature are painted. Hardwoods due to the figurative grain present in their structure are usually stained and clear coatings such as varnishes are applied.

Q3   The natural defect present in softwoods is the 'knot', which exudes copious amounts of resin. This resin can prevent oil based paints applied over the knot from drying. If the knot is not sealed the resin can bleed through and discolour the applied paint system.

Q4   The substance applied as a sealer over the resin is named 'knotting'. It is a mixture of shellac resin and methylated spirit.

Q5   Timber should always be abraded along the grain, never across the grain.

Q6   Softwoods include cedar, fir, pine and spruce. Hardwoods include oak, ash, beech, sycamore, mahogany, alder, teak.

Test your knowledge 2.3

Q1 Steel
Q2 Iron
Q3 above 80%
Q4 75 microns
Q5 Flame cleaning

Test your knowledge 2.4

Q1 Plasters and brickwork are porous surfaces by nature and will absorb or suck in applied paint coatings if a sealer is not applied to the surface first. A sealer can be a diluted first coat of the paint system to be applied or special sealers can be obtained from the retailer.

Q2 New plaster surfaces are never abraded prior to the application of coatings as the surface could become scored by the action of the abrasive and the particle size on the abrasive. Scores of scratch marks could be created and these marks will show through applied coatings. Filling will then be required to hide such scratching or scoring.

Q3 Alkali deposits that may be present in new plaster can be detected by using a solution of universal indicator or litmus paper on the surface. If the solution or the paper turns blue it indicates the presence of free alkali.

Q4 If salts are present in brickwork or plaster, they can be seen as a white powder on the surface. The salt is brought to the surface and deposited there when the water in the material evaporates. This salt must be brushed off dry before any paint system is applied. The defect or presence of salts is known as 'efflorescence'. If oil based paint is applied on a surface where there is still moisture and salts present the salts attack the oil medium in the paint and turn it into a brown sticky solution known as 'saponification'.

Test your knowledge 2.5

Q1 To remove wallpapers by hand you will require a bucket, hand scraper and a supply of hot water. Remove all items from the walls to be stripped. Protect all surrounding surfaces and then make a test to determine method of removal. If the existing paper is vinyl, it should be possible to peel off the top layer leaving a background paper. This paper can then be soaked with hot water at least two to three times. It should then be removable using the scraper. Bag all rubbish and dispose of in the correct manner.

   If the paper does not move, soak each wall until the paper can be easily scraped off the surface. If the paper proves resistant to scraping, apply another coat of hot water until it can be easily removed. If the paper does not come away from the wall, a steam stripper may have to be employed.

Q2 The illustration is an electrically operated steam stripper. The steam stripper must be checked to see if it has any visible faults such as frayed wires or damaged connections. Check that the portable appliance testing label is current. Assemble the equipment by attaching the hose to the hose outlet on the main body of the stripper. Remove filler cap and fill $\frac{3}{4}$ full with hot water. Replace the filler cap. Connect to the power supply to allow the water to boil and produce steam. When steam is produced place the condenser plate against the wall to soften the wallpaper which can then be removed using a scraper. Never point the condenser plate towards colleagues when working with such equipment. Be aware that scalding could take place.

Q3 Steam strippers should be connected to 110-v extension cables attached to 110-v transformers. The transformer should be connected to a 240-v power source. Good practice indicates that the shortest cable supply of 240 v should be from the socket to the transformer.

Test your knowledge 2.6

Q1 Wet and dry abrasive paper is known as 'silicone carbide'. Two lubricants that can be used with this abrasive are water or turpentine.

Q2 If it is suspected that the paint system to be prepared is old, the indications are that it will contain lead, possibly in the form of the primer or in the enamel coating. From the 1970 onwards little lead was used in paints. However the wet process of abrading should be adopted if lead is suspected.

Q3 Abrasives can be purchased in:

- sheet
- roll
- as discs or belts
- as powders
- in block form.

Q4 Portable appliances can be powered by:

- electricity
- compressed air
- battery.

Q5 The grading codes used on dry abrasive products is coded from 00 flour, very fine to 3 coarse. The lower the number the finer the preparation.

Q6 The function of an abrasive is to wear away a surface to remove undulations or brush marks. To remove peaks and troughs till it become flat smooth and can receive a paint film.

Q7 Portable sanders include:

- palm sanders
- orbital sanders
- disc sanders
- rotary sanders
- belt sanders.

Q8 Personal protective equipment that must be worn when using portable sanding appliances are:

- respirators or dust masks
- safety spectacles or goggles
- overalls.

Q9 The function of a transformer in a 110-v set up is to reduce the normal 240-v supply to 110 v thus reducing the chance of fatality when using electrical equipment.

Test your knowledge 2.7

Q1 The illustrations shown indicate:

- regulator with pressure dial
- snap connectors for quick hose connection and quick release.
- dual connector
- fuel bottle.

Q2 To reduce the risk of electrocution when setting up hot air equipment use a 110-v transformer and 110-v extension cable and plugs.

Q3 To ignite LPG equipment follow these procedures:

- connect regulator to supply bottle
- connect hose to regulator
- connect torch to hose
- turn on the bottle to allow gas to the regulator
- adjust the regulator to allow gas into the hose
- turn on the gun control to allow gas to escape and ignite the gas with a spark gun
- adjust the flame to the correct burning pressure.

To extinguish the torch

- turn off the gas supply at the bottle
- let the gas burn out
- turn off the regulator and torch control
- coil up the hoses and store the equipment

Q4 Two hand tools to use when removing paint coatings from surfaces are:

1. Scraper or broad knife – used to remove paint from flat areas after heat has been applied and the coating has softened.
2. Shave hook – combination is used to remove paint from moldings after the coating has softened.

Q5 The paint removal process should cease at least one hour prior to leaving premises. This is to check for the possibility of smoldering embers bursting into flame and creating a serious fire hazard.

Q6 The sign or label applied to the housing of a hot air gun would be the portable appliance test date.

Q7 Liquid propane bottles should be stored upright in compounds situated outside. These compounds should be fenced and appropriate signage displayed.

Q8 Advantage – LPG equipment for paint removal is an economic method as the fuel is cheap.
Disadvantage – the harshness of the flame can lead to scorching of timber when used by unskilled operators.

Q9 Advantage – Hot air guns are excellent for removing paint coatings from surfaces located internally.
Disadvantage – when used externally the heat generated by the gun element is quickly dispersed thus making the use of the gun ineffective.

Test your knowledge 2.8

Q1   Information is placed on containers to provide the user with relevant facts relating to the contents. Information such as:

- manufacturer's instructions
- drying times
- application methods
- health and safety or COSHH information.

Q2   Personal protective equipment (PPE) that should be worn when using paint removers is:

- protective overall and apron
- safety spectacles or goggles
- hand protection such as gloves
- long sleeved clothing to cover the arms
- safety footwear
- respirators.

Q3   Paint removers if inhaled can cause respiratory problems, they are also cancer forming and burn the skin.

Q4   The advantages of removing paint coatings by the use of liquid strippers are as follows:

- enables removal of paint from moldings
- enables removal of coatings from metals without damage
- enables removal of coatings next to glass.

Chapter 3

Test your knowledge 3.1

Q1   Items that will aid the preparation of coatings are:

- Container opener
- Paint stirrer or attritor
- Sieve or strainer
- Paint container such as a kettle, bucket, trough or tray
- Palette knife
- Brush

Q2   Paints and materials require preparing for use to enable the product to meet performance requirements. Pigment and oil separate when paint is stored for long periods and require remixing before use. Adhesives require water to be added to dry crystals to activate the adhesive and turn it into a glutinous solution. The products can be checked for transport damage, manufacture damage or frost damage.

Q3   The information available on manufacturer's instruction labels or leaflets is varied and numerous. Listed are some possibilities:

- technical information on how to use and apply that product
- special instructions
- COSHH information
- HASWA information
- safety signs/warning signs

- coverage
- drying times
- recommended preparation
- application methods
- shade and batch numbers on wallpapers.

Test your knowledge 3.2

Q1 Fillers can be purchased as:

- Powders that have to be mixed with water to a smooth paste and then be applied to specified areas.
- In caulking tubes that are mastics and have a plastic nature. These fillers are applied using mastic guns and surplus filler should be removed from surrounding areas after the crack or indentation has been filled.
- Ready mixed in tubs. Various grades from fine surface to coarse grain can be purchased for use.

Q2 The illustrated item is a sieve. It is used to strain dirt and grit from previously used paint coatings. Skins that form on the surface of the paint inside the container can also be collected by the sieve during straining thus reducing surface contamination whilst using previously used products.

Q3 Three containers that the painter dispenses paint coatings into ready for use are the:

1. *Painter's kettle or pot* – Oil based primers, undercoats and finishing coats are held in the kettle during application. By pouring a required amount of the coating from the stock container into the kettle it reduces the risk of contaminating the whole stock whilst using the paint.
2. *Bucket* – Used to decant water based paints which enables large brushes to be used to apply the coating to surfaces.
3. *Roller trays or buckets* – These containers hold large quantities of material for application to surfaces using rollers.

Q4 Prior to unwrapping the packaging from wallpaper check the:

- batch number
- shade number
- roll number
- check for visible roll end damage
- check for signs of manufacture or carrier damage.

Chapter 4

Test your knowledge 4.1

Q1 The three documents used to obtain materials to paying for them are the:

- order form
- delivery note
- invoice.

Q2   Precautions to take to protect the public from hazard and risk during painting operations can be as follows:

- restrict access to the immediate working area
- position warning signs
- use protective sheeting and tarpaulins to prevent splashing or falling objects
- post information notices on dates of operations
- inform the council if working on highways to place prohibition parking signs or cones
- construct pedestrian walkways.

Q3   Adverse weather conditions can cause delays or halt painting operations if it is too windy or blustery to safely work from scaffolding. Early morning fog or sea fret can soak external surfaces which are to be painted. Paint coatings can be spoiled by dust, water or moisture contamination after application of paint systems due to sudden weather changes.

Test your knowledge 4.2

Q1   Small items removed from locations prior to decoration should be generously wrapped and stored in containers to protect them from damage. The containers should be marked to identify contents for later replacement. Door, window fittings and fixtures should be stored together to prevent misplacement of screws or brackets.

Q2   Prior to removing electrical fittings turn off the power supply at the distribution box and remove the fuse. Place a notice on the box to inform other persons of your activities. As soon as you have completed the task replace the fuse and restore the power.

Q3   When removing curtains and blinds:

- Curtains – unhook the retaining hooks from the rail, noting the configuration or placement of the hooks to enable replacement after decoration. Fold the curtains in pleats and lay out in a spare room to prevent creasing.
- Blinds – close the blinds, locate and open the retaining brackets and remove the housing and lats. Lay it flat in a safe location. Observe the correct way to reinsert the fitting to the brackets after decoration to ensure the blind will open and close correctly.

Q4   Wooden strip floors should be sheeted up and taped to protect the laminate. Brown paper, polythene or cotton dustsheets can be used. Be aware that polythene offers a slippery surface on which to work.

Test your knowledge 4.3

Q1   The persons to whom you may be accountable to at your place of work could include:

- client or clientele
- employer or representative such as the foreperson or charge hand
- immediate superior
- clerk of works
- site agent

- general foreperson
- health and safety executive inspector.

Q2 Paperwork that could assist in the carrying out of operations could include:

- memo
- specification
- schedules
- codes of practice.

Q3 To maintain customer satisfaction on completion of a job consider the following:

- check for paint splashes and remove
- clean windows
- hoover carpets
- dispose of debris, do not fill the customers bin
- check for accidental damage
- ask the customer if they are satisfied with the work
- meet any reasonable request related to the work carried out.

Chapter 5

Test your knowledge 5.1

Q1 The part of a paint brush that has a natural taper, curve, fiscules and a flagged end is the 'bristle'.

Q2 The purpose of fiscules, serrations and flagged end of a bristle is to hold the paint when it is loaded on to the brush to the point when it is placed on the surface where it is distributed.

Q3 The four parts of the paint brush are the:

1. Filling – the bristles of the brush.
2. Setting – the glue that holds all the bristles in place to the shape required.
3. Ferrule – the metal support structure that holds the filling and brush handle together.
4. The handle – usually constructed from one of the following timbers, ash, alder or beech.

Test your knowledge 5.2

Q1 Any acceptable drawing.

Q2 The function of a radiator roller is to coat the reverse side of the radiator with the specified paint coating without the need to remove the radiator.

Q3 Brushes and rollers can be stored during short breaks such as lunches and overnight:

- wrapped in airtight polythene bags
- in vapour keeps
- suspended in water
- stood in the paint for very short breaks.

Q4   Roller sleeve materials are as follows:

* Foam or sponge – used to apply gloss paints.
* Mohair – used to apply eggshell or gloss paints.
* Lambs wool – used to apply emulsions and water based coatings.
* Synthetic – used to apply water based coatings.

Roller sleeves are available in short, medium or long pile to enable the application of coatings to surfaces of light to heavy texture.

Chapter 6

Test your knowledge 6.1

Q1   The minimum recommended thickness of a three-coat paint system is 75 microns.

Q2   The four reasons for applying paint coatings to surfaces are for:

1. decoration
2. identification
3. hygiene
4. protection.

Q3   The function of a paint system is to offer protection to building materials from natural or man-made influences. For detailed information refer to Section 6.5.

Q4   The three main constituents of paint are:

1. Pigment – the solid colouring matter.
2. Medium – the liquid part of the paint, usually an oil or resin or combination of both. Sometimes referred to as the binder or the vehicle.
3. Solvent or thinner – enables the paint to become workable.

Q5   The main drying mechanisms of paint are:

* Evaporation
* Oxidation
* Polymerization
* Coalescence

Test your knowledge 6.2

Q1   The varnish most suited to environments where excessive moisture or water is present is a 'long oil' varnish which is very elastic and the applied film can expand and contract with the surface.

Q2   Two-pack varnishes have excellent resistance to chemical and moisture attack.

Q3   Pot life is a term used to describe how long a paint remains in its liquid state after the activator has been added to the base material.

Q4   Decorative stains enhance and colour the surfaces of timbers. They define the grain patterns of hardwoods.

Q5   An ideal stain to use as a protective coating on exterior timber would be the micro porous types on good woodwork or the preservative type for rough timber in the garden.

Chapter 7

Test your knowledge 7.1

Q1 The laying off process of paint applied to a flat surface is:

- to spread the paint to an even consistency
- to brush in the shortest direction applying slight pressure and over-lapping each brush stroke by a third of the brush bristle width
- work along the surface in this manner in areas of approximately 300 mm × 300 mm
- do not load any more paint on to the brush
- lay off in the long direction of the surface with lighter brushstrokes to eliminate brush marks, again overlapping each brushstroke by a third
- slight pressure will be needed to work each applied stage into each other creating the paint merge.

Q2 The procedure for painting a panelled door is:

- mouldings first, then panels the top two panels first
- working down the door paint the top rail and top muntin
- next paint the top intermediate rail
- continue with the mouldings and panels of the middle door
- paint the middle muntin
- paint the mouldings and panels at the base of the door
- paint the middle rail, muntin and bottom rail
- paint the hinge stile
- paint the latch stile and finally the edge of the door if it is the inside face in the room
- if the door is the outside face of the room, the hinge edge would be painted before the hinge stile.

Q3 Defects that can occur during the application of paint to a door are:

- excessive brush marks
- ladders caused by incorrect laying off
- sags and runs caused by overloading of the coating
- grinning due to over brushing of the coating
- dry edge marks caused by slow operator application of the paint coating.

Q4 The term 'wet edge' refers to the applied coating at the point where one working application of paint meets the next working application of paint. The first edge will have dried slightly and the second joining edge will have to be worked into the first by applying more pressure to the brush. This will enable the flow to occur and join the two edges of the paint with an invisible join. If this operation is not carried out correctly, the edge can be seen as a fat edge build up and is unsightly.

Q5 Four linear components of a room that may require paint to be applied to their surfaces are:

- Skirting
- Dado rail
- Picture rail
- Cornice or coving
- Panel mouldings
- Architraves

Chapter 8

Test your knowledge 8.1

Q1   A pattern book is a compilation of the wallpapers that a manufacturer or a retailer offers for sale to a customer. It allows the customer to browse and choose, then order.

Q2   Preparatory papers will include:

- Lining paper of various grades for the purpose of providing even clean surfaces on which to apply wallpapers.
- Reinforced lining papers used to mask poor surface irregularities such as deep cracks.
- Foils used to provide a barrier against penetrating dampness.
- Expanded polystyrene used as an insulator.

Q3   RD indicates that the wallpaper is relief decoration. It has an emboss or is raised from the surface.

Q4   Linen backed lining paper should be used on badly cracked wall surfaces.

Q5   Vinyl and blown vinyl wallpaper can be removed from the surface in the dry state.

Q6   Selvedge is the protective strip of paper that prevents damage to rolls of expensive wall hangings. Information such as the designer and colourways are printed on selvedges.

Q7   Grades of lining paper available for use are 800, 1000, 1200 and 1400.

Test your knowledge 8.2

Q1   Prior to unwrapping rolls of wallpaper check that the shade and batch numbers are the same. Check for visual damage to the roll especially the roll ends.

Q2   A drop pattern can be identified if the pattern on the left edge of the paper is different from the pattern on the right edge of the paper and will not match up. If an imaginary line were drawn horizontally from left to right edges, the two points should match. To check use two rolls. Identify a prominent feature on the pattern on both rolls and place them side by side. If one roll has to be moved up 265 mm it will indicate a drop match.

Q3   The ideal starting point for commencing the hanging of wallpaper in a room is from a light source to the entrance most used, or start from the furthest point in the room from the entrance and work towards the entrance. All rooms are individual and the starting point has to be determined relating to the number of windows and doors in the room.

Q4   The first length of wallpaper hung on a staircase should be the longest length in the stairwell.

Q5   To ensure that patterned paper is centered on a wall or ceiling find the middle point of the surface and strike chalk or plumb lines. Hang the paper from these lines and work both directions until complete. To centre pattern design, measure the repeat and divide the pattern repeat into the height and width of the wall or ceiling. This will enable you to position the patterned paper to best effect on the surface.

Q6   The standard dimension of a roll of English wallpaper is 10.05 m long by 0.530 m wide.

Q7   When turning corners check the plumb of the corner. Internal corners require the paper to be fitted to the corner with a minimum overlap around the corner. The remaining piece of the width of wallpaper is then applied to the next wall and plumbed. External corners, if true, can be papered around the angle without cutting, however if it is not plumb the surplus requires feathering and the next piece requires matching and fitting up to the edge and trimming before continuing.

Test your knowledge 8.3

Q1   Fungicide is added to cold water adhesives to prevent the development of mould and fungi under applied vinyl wallpapers. The fungicide is incorporated during the manufacture and packaging process.

Q2   The process or method of folding pasted lengths of wallpaper which is to be hung horizontally is the concertina method. Make small folds of about 300 mm which are easily manageable during application.

Q3   The tools materials and equipment required to measure, cut, paste and fold wallpaper are:

- ruler or tape
- shears or trimming knife
- straightedge and trimming knife
- ridgley straightedge and trimmer
- pasting brush
- bucket and adhesive
- protective covers
- cloths and sponges
- paperhanging bench or table.

Test your knowledge 8.4

Q1   The trimming of wallpaper should be carried out using new blades in trimming knives and sharp shears or trimming wheels to ensure an accurate and clean cut.

Q2   To trim waste paper around light sockets:

- first turn off the power at the distribution box and isolate fuse
- post a warning notice
- loosen the socket
- make a star trim to the edge of the socket plate from the centre or four cuts from the centre of the plate to the corners of the plate
- trim off the waste, leaving a selvedge which will tuck under the socket plate
- relocate and tighten the socket plate
- clean off paste from the plate
- replace fuse, turn on the power and remove the warning sign.

Q3   Metal straightedges should be used with trimming knives as timber or plastic devices can be damaged by the trimming knife blade.

Q4   The difficulty encountered when trimming around window reveals is loosing the plumb of the wallpaper and all the individual pieces required to complete the processes. The pattern continuity must be maintained with no visible errors.

Q5   All trimming tools and equipment used during the paperhanging process should be frequently cleaned using warm water and cloth to remove build up of wet or dry adhesive. The items should also be dried before storage.

Chapter 9

Test your knowledge 9.1

Q1   The natural order of colour is the organising of the phenomenon of colour when white light is broken up into its parts. Thus the spectrum as we know it is in a band where the colours change and merge into each other in an order. This order is most useful when the lightest colour yellow is followed through the spectrum till yellow is again reached. The colour circle or wheel is often referred to as the natural order of colour.

Q2   Colours that cannot be mixed from other colours are termed primary colours. Pigment primary colours are red, yellow and blue. Light primary colours are red, green and blue.

Q3   Secondary colours are produced by the mixing two primary colours together, thus:

- mixing red and yellow together would produce orange
- mixing yellow and blue together would produce green
- mixing blue and red together would produce purple.

Tertiary colours are produced by mixing two secondary colours together, however in some textbooks it states that a primary and secondary colour mixed together will produce a tertiary colour. The three tertiary colours are olive, russet and slate.

Q4   White light is separated into all its parts when one ray of light is passed through a prism.

Q5   When black is added to a colour the colour is termed a shade of its true colour.

Q6   When white is added to a colour the new colour is a tint of its original colour.

Q7   If black and white is added to a single colour the term applied to the produced colour is monochromatic.

Q8   The framework of BS 4800 is in three parts:

1. horizontal rows of colour numbered from 02 to 24
2. sections of colours with grayness from A to E
3. vertical rows of colour with weight numbered from 01 to 55.

There are some variations to this general rule.

Q9   The three parts of the BS 4800 colour code are:

1. Hue – even number
2. Greyness – letter from A to E
3. Weight – odd number from 01 to 55.

08 B 15 denotes a buff or light cream colour often known as magnolia.

Q10   The leaflet that can be obtained at no cost from the retailer is the BS 4800 colour card.

Chapter 10

Test your knowledge 10.1

Q1   The four categories of decorative effect can be termed:

1. broken colour
2. graining
3. marbling
4. stencil work.

Q2   Oil based eggshell is one of the best recommended paints to use as a base coat ready to receive decorative treatments. It allows applied media to flow on its surface whilst being manipulated into the required treatment.

Q3   Gilp is the wetting agent used on surfaces prior to the application of media for the production of marbling effects. It is mixed from equal parts of turpentine and linseed oil with a few drops of added driers.

Q4   The two media that can be used to apply decorative effects are:

1. water based
2. oil based.

Q5   Sponging and rag rolling can be applied to a surface without using brushes to create the effect.

Q6   The name given to the part that holds the stencil together is the 'tie'. The tie should appear as an integrated part of the stencil.

Q7   The effects that can be produced during graining are:

- dragging
- flogging
- combing
- figure work
- over graining
- brush graining.

Q8   Materials that can be used to produce stencil plates are:

- Mylar
- acetate sheet
- oiled paper
- shellac coated paper
- zinc sheet
- card.

Q9   The brushes, tools and materials that could be used to produce a piece of marble are:

- palette
- dipper
- palette knife
- stainers or tinters
- glaze or scumble
- turpentine
- selected paints
- varnish
- application or rubbing-in brush
- cloth usually lint free

- fitches and filberts
- softeners
- feathers
- sign writing pencils and feathers.

Q10 The process of producing the rag rolled effect is as follows:

- apply the mixed glaze to the surface and hair stipple
- roll up a lint free rag and randomly roll over the whole surface or
- mix up a thin paint and soak a cloth with the paint
- squeeze out surplus paint and roll the rag over the surface.

The adding of paint to the surface with this treatment is known as ragging on and is usually carried out using water based paints. The removing of paint from a surface is known as ragging off and is carried out using oil based paints.

Chapter 11

Test your knowledge 11.1

Q1 The ratio of a ladder against a wall when erected should be 4:1. That is four up, one out. The angle at the base of the ladder to the ground should be 75°.

Q2 A ladder should project five rungs above a working platform, a distance of 1.070 m minimum.

Q3 Precautions to take when using ladders:

- raise to the correct ratio
- secure the ladder before use at the top, bottom or by other persons footing the ladder
- check for defects to the ladder
- ensure the ladder is clean and your footwear is not contaminated excessively with ground earth
- do not over reach
- do not jump the ladder.

Q4 Prior to using stepladders check:

- ropes and stays are in perfect condition
- all screws and bolts are tight
- there is no visible damage to the stepladder
- that the stepladder is of a suitable size for its purpose.

Q5 Tie rods are fitted under stepladder treads. The more ties the better the quality of the stepladder.

Q6 The distance between rungs on an extension ladder is 230 to 250 mm.

Q7 Paint should not be used as a protective coating on ladders as it will mask defects such as splits or cracks.

Q8 The minimum rung overlap on a ladder extending up to 4.8 m should be two rungs.

Q9 The device that prevents a trestle from collapsing is the locking hinge.

Q10 Scaffold board ends are protected from damage by attaching a zinc strip, or by chamfering the corners.

Q11 50-mm thick scaffold boards must be supported every 2.5 m.

Q12 The maximum overhang of a scaffold board is four times the thickness. The minimum overhang is 50 mm.

Chapter 12

Test your knowledge 12.1

Q1 The organisation responsible for enforcing safety law is the 'Health and Safety Executive'.

Q2 The health and safety executive can:

- entering of premises at any reasonable time (day or night) if a dangerous situation is occurring
- examination and investigation of situations as deemed necessary
- collection of evidence either manually or electronically
- examination of relevant documents
- making safe of dangerous equipment
- questioning of persons and declarations of truth to be signed
- use of the police force to effect entry to premises.

Q3 The employer's responsibilities:

- Provide and maintain plant, machinery, equipment and ensure safe systems of work are adhered to.
- Provide training, give instruction, produce information and implement supervision in the workplace to maintain the health and safety of employees.
- Provide and maintain a safe place of work without risks to health and welfare.
- Set up procedures for handling, storage and transport of articles and substances in compliance with HASWA.

Q4 The consequences of accidents
The implications to the employer:

- the loss of labour
- insurance premiums could rise
- faces prosecution or civil action
- worry and stress
- loss of business
- gains a bad reputation.

The implications to the victim:

- loss of life
- enduring pain and suffering
- can suffer a disability
- loss of livelihood
- financial loss
- prosecution.

Q5 HASWA is the Health and Safety at Work Act.

Test your knowledge 12.2

Q1 The COSHH regulations were introduced by the Health and Safety Executive.

Q2 The purpose of the regulations was to protect people in the workplace from the effects of substances hazardous to health.

Q3 Regulations relating to COSHH that are the employer's and employee's responsibility
Employers duties

- assess the risks to people's health from the way the substance is used in the workplace
- prevent anyone from being exposed to the substance
- decide how to control exposure to reduce the risk
- establish effective controls
- train and inform the work force.

Additional duties
The employer needs to monitor employee's health and exposure to such substances.
Employee's duties in identifying hazards
The employee needs to know more than the fact that a hazard exists. The employee should be trained to understand these hazards and risks when using such substances.
The employee has to know:

- what the risks are from using such substances in the workplace
- how the risks are controlled
- what precautions have to be taken when using such substances.

Q4 A hazardous activity carried out by a painter and decorator that could cause harm would be the use of paint removers in unventilated spaces. Any reasonable answer would be considered as evidence.

Q5 The three materials that have their own regulations are:

1. lead
2. asbestos
3. radioactive material.

Q6 Personal protective equipment that could be worn to lessen the risk of contamination could include:

- Dust masks/respirators
- Goggles
- Gloves/overalls/safety footwear.

Q7 Records to be kept by employers for reference could include:

- staff details
- accident book
- scaffold log
- maintenance log
- PAT testing log (portable appliances)
- others

Q8 The implementation of safe working practice in the workplace lies with the employer and employee.

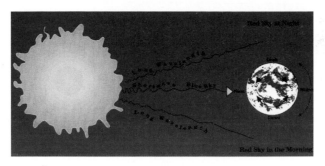

Plate 1 *Light waves from the sun to the earth (see Figure 9.3)*

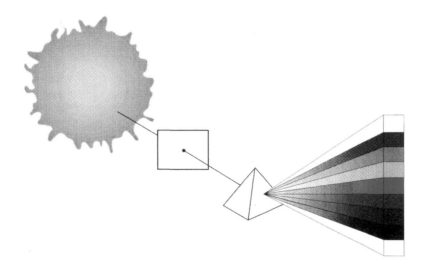

Plate 2 *Creation of the spectrum (see Figure 9.4)*

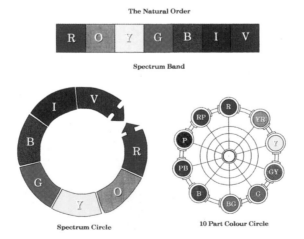

Plate 3 *The natural order and colour wheel (see Figure 9.5)*

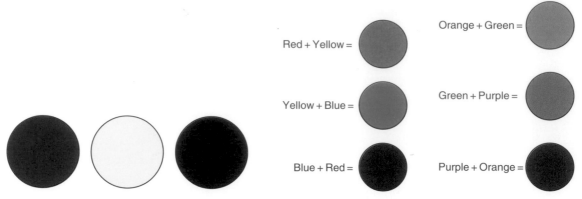

Plate 4 *Pigment primary colours (see Figure 9.6)*

Red + Yellow =

Yellow + Blue =

Blue + Red =

Plate 5 *Secondary colours mixed from primary colours (see Figure 9.7)*

Orange + Green =

Green + Purple =

Purple + Orange =

Plate 6 *Tertiary colours mixed from secondary colours (see Figure 9.8)*

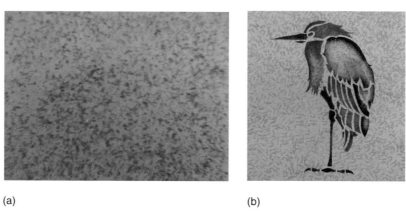

(a)

(b)

Plate 7 *Sponge stippling (see Figure 10.15)*

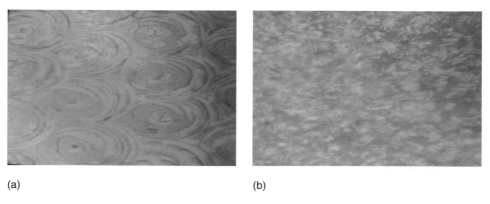

(a)

(b)

Plate 8 *Rubber stippling (see Figure 10.16)*

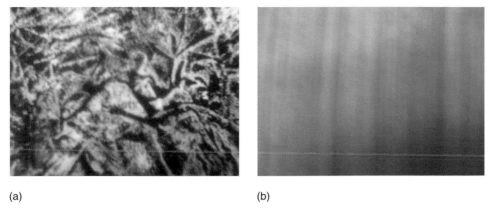

(a)   (b)

Plate 9 *Rag rolling and dragging (see Figure 10.17)*

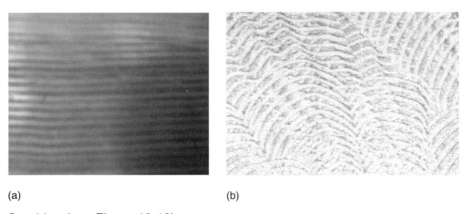

(a)   (b)

Plate 10 *Combing (see Figure 10.18)*

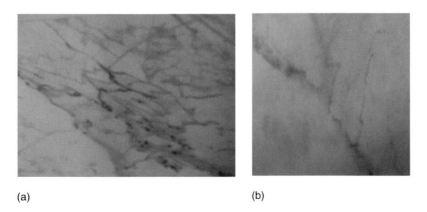

(a)   (b)

Plate 11 *Carerra marble examples (see Figure 10.19)*

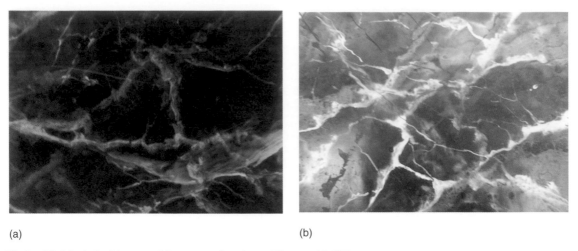

(a)   (b)

Plate 12 *Vert de Mer marble examples (see Figure 10.20)*

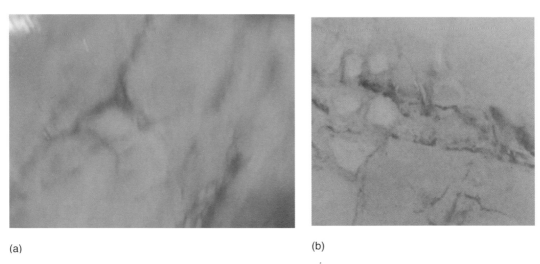

(a)   (b)

Plate 13 *Sienna marble examples (see Figure 10.21)*

Plate 14 *Gilders cushion, knife and tip (see Figure 10.27)*

Index

Abrading by hand, 40
Abrasive paper, 38
Accent, 158
Access scaffold, 183
Achromatic colour, 158
Acrylic primer undercoat, 109
Adhesive, 134
Adjoining property, 76
Advancing colour, 158
Alkyd gloss, 111
Aluminium ladders, 184
Aluminium wood primer, 109
Anaglypta, 130
Analogous colour, 158
Application of adhesive, 143
Application of materials, 116
Applied paint system, 116
Applied paintwork, 118
Architrave, 120
Asbestos, 199
Assess metal surfaces, 23

B.S.4800 paint colours for building purposes, 159
Badger softeners, 165
Basic adhesives, 133
Basic tools, 5
Batch number, 136
Batons, boards and staging, 189
Bleeding of the resin, 22
Blistering, 205
Blooming, 205
Blown vinyl, 132
Border adhesives, 135
Broad knife, 3
Broken colour, 164
Brush construction, 91
Brush keeps, 14
Building boards, 22

Carerra, 174
Cartridge cases, 65
Caulking tool, 65
Caustic paint remover, 55
Cellular structure of hardwoods, 21
Cellulose paste, 134
Chalking, 205
Chisel knife, 4
Cissing, 205
Coalescence, 105
Coastal environments, 18
Coatings to ceilings, 118
Cold water cellulose pastes, 134
Colour and light, 154
Colour card, 160
Colour circle, 158
Colour codes, 160
Communication, 74
Comparisons between paint removers, 55
Comparisons of heat removal equipment, 45
Concertina fold, 142
Condenser plate, 34
Coniferous, 19
Constituents of paint, 103
Construction of brushes, 91
Contrasting colours, 158
Corrosion, 25
COSHH technical data, 199
Coving or cornice, 125
Crypto-crystalline salts, 28
Curtains, curtain rails, 84
Customer satisfaction, 72
Cutters, 166
Cutting and pasting wallpaper, 141
Cutting and trimming, 151

Dado, 125
Dado rail, 125

Decayed and denatured wood, 22
Deciduous, 19
Decorative paint effects, 163
Decorative papers, 130
Decorative plasterwork, 27
Decorative stains, 113
Defective plaster, 28
Deliquescent salts, 28
Dermatitis, 200
Discordant colour, 158
Double arm rollers, 94
Dragging, 172
Driers, 103
Drop pattern, 136
Dry abrading, 43
Dry-out, 28
Drying of coatings, 104
Dust brush, 7
Dust inhalation, 199
Dust sheets, 87
Duties and responsibilities, 194

Efflorescence, 28
Eggshell, 110
Employee's awareness, 75
Employee's duties, 195
Employer's duties, 194
Emulsion paint, 112
End grain, 21
Erecting, lifting and carrying trestle scaffolds, 189
Evaporation rate, 104
Expanded polystyrene, 129
Extenders, 103
Extension cables, 14
Extension ladders, 184

Fan fitch, 165
Felt roller, 9
Ferrous metals, 23
Ferrule, 93
Filling, 66, 92
Fire extinguishers, 47
Fitches, 165
Flame cleaning, 25
Flogger, 166
Flogging, 177

Gauzes, 63
Gilding, 164
Gilp, 174
Gold leaf, 180
Good working practice, 145
Graining, 164
Greyness, 159
Grills or ventilation covers, 82
Ground coats, 169

Hair stippler, 166
Hammers, 5
HASWA, 44
Hazardous materials, 199
Hot air guns, 47, 51
Hue, 159

Impact adhesives, 135
Ingrain, 130
Inspections and investigations, 196
Ironmongery, 82

Job sheet, 81
Juxtaposition, 158

Kettles, 7, 59
Knives, 8
Knotting, 22

Ladders, 183
Lambs wool roller, 95
Lap fold, 142
Lead, 199
Lifting and carrying ladders, 185
Lincrusta, 132
Lincrusta glue, 135
Lining paper, 128
Liquefied petroleum gas, 49
Liquid stripper, 54

Mahogany graining, 177
Marbling, 164
Masking paper and tape, 88

Mechanics of corrosion, 25
Metal foils, 129
Mill scale, 23, 25
Moisture content, 20
Monochrome, 158
Mottler, 166

Nail punch, 6
Natural fibres, 92
Natural order of colour, 158
Non-porous surfaces, 29

Oak graining, 177
Orbital sanders, 13, 44
Oxidation, 104

Paint removers, 54
Paint schedule, 80
Paint system, 106
Palette knives, 4
Palm sander, 13
Panels, 52, 120
Paperhanging brush, 7
Paste table, 11
Pasting brush, 10
Pasting machine, 13, 14
Pigment, 103
Plaster surfaces, 26
Plasticizer, 103
Plumb bob, 9
Polymerization, 104
Portable appliances, 41
Pot hook, 3
PPE, 44
Preparation of surfaces, 17
Primary colour, 156
Protection of areas, 86
Pure bristle, 92
PVA adhesive, 135

Radiator rollers, 95
Radiators, 85
Rag rolling, 171
Raising and lowering ladders, 184
Reasons for painting, 102

Reinforced lining paper, 129
Reinstatement of work area, 89
Relief vinyls, 131
Ridgley trimmer, 10, 146
Roller construction, 94
Rollers, 95
Rulers, 8
Rural environments, 18

Saponification, 28
Selection of brushes, 98
Settings, 93
Shave hooks, 4
Shears or scissors, 9
Shot blasting, 25
Solvent, 103
Solvent remover advantages, 55
Solvent remover disadvantages, 55
Spirit levels, 9
Sponges, 167
Stain, 113
Steam strippers, 34
Stencil knife, 168
Stencil plates, 178
Stencil work, 164
Stock, 93
Straight edge, 146
Supaglypta, 130

Tapes, 8
Tarpaulins, 87
Tertiary colours, 157
Theory of colour, 154
Tint, 159
Top rails, 120
Transformer, 14
Trestles, 188
Trimming knives, 8
Trimming procedures, 146
Types of abrasive, 37
Types of brush, 97
Types of roller, 94

Vacuum cleaner, 14
Vandalism, 17
Varnishes, 112

Vert de Mer, 175
Vinyl, 132

Washing down, 18
Waterproof abrasive papers, 38

Wear and tear, 17
Weight, 186
Wet abrading, 42
Wet edge time, 116
Wire brushes, 6